William Henry Fowler

Fifty years' history of the development of Green's Economiser

William Henry Fowler

Fifty years' history of the development of Green's Economiser

ISBN/EAN: 9783337204341

Printed in Europe, USA, Canada, Australia, Japan

Cover: Foto ©ninafisch / pixelio.de

More available books at **www.hansebooks.com**

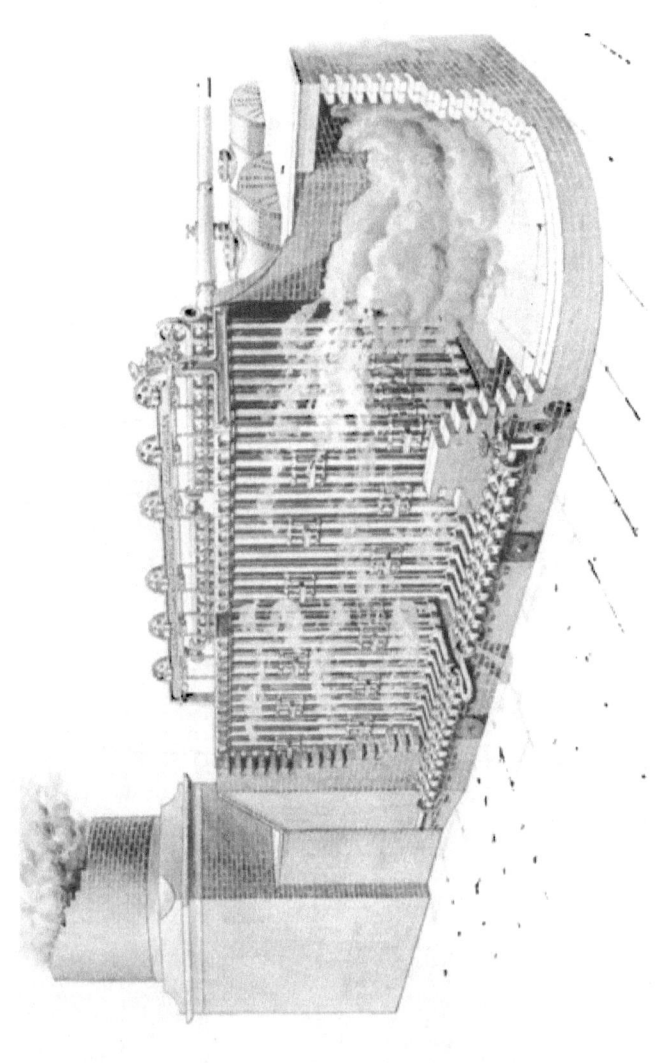

E. GREEN & SON LTD.

Sole Makers and Original Inventors of

GREEN'S PATENT
FUEL ECONOMISER

2 EXCHANGE ST. MANCHESTER

WORKS
WAKEFIELD YORKSHIRE ENGLAND
AND
MATTEAWAN N.Y. U.S.A.

Telegraphic Addresses	Telephone Nos.	
"ECONOMISER MANCHESTER"	MANCHESTER	No. 1375
"ECONOMISER WAKEFIELD"	WAKEFIELD	No. 43

AND AT
LONDON GLASGOW BIRMINGHAM BRUSSELS MULHOUSE
ROUEN MOSCOW VIENNA & JOHANNESBURG.

LIST OF CONTENTS

	PAGE
INTRODUCTION	1-3
THE INVENTOR OF THE ECONOMISER	4-6
Biographical Sketch of the late Mr. Edward Green.	
THE FIRST EXPERIMENTAL ECONOMISER	7-9
THE DEVELOPMENT OF GREEN'S ECONOMISER	10-34
The first Economiser Patent, Scrapers for Economiser Tubes, Driving and Reversing Gear for Scrapers, Lids and Seatings for Economiser Tubes, Sectional Top and Bottom Boxes, Economiser Bottom Boxes, Economiser with Feed Downflow.	
OTHER ECONOMISER INVENTIONS	35-70
Lees' Steam Brush, Whitehead's Steam Brush, Twibill's Circulator, Mannock's Scraper, Robinson's Scraper, Whitehead's Chain Scraper, Holt & Galloway's Continuous Economiser, Needham's Scraper, Screw Shaft and Reversing Gear, Obach's Tube Joints, Reversing Gear and Scrapers, Bell's Spiral Economiser, Mundy's Scraper, Calvert & Taylor's Continuous Economiser, Elson's Bottle Economiser, Lees & Garforth's Scraper and Continuous Economiser, Twibill's Circular Economiser, Mason & Alcock's Continuous Economiser, Prestwich & Pimbley's Split-tube Economiser, Bell's Continuous Economiser, Sykes' Group Economiser, Lowcock & Taylor's Scrapers, Perkins and Scott's Scrapers and Lids, Lowcock & Sykes' Blow-out Pipes, Bell's Continuous Economiser, Hawkin's Top and Bottom Boxes, Lowcock & Sykes' Bottom Boxes, Twibill's Internal Tube Economiser, Sykes' Economiser Lid, Burpee's Spigot Joints and Tapered Tubes, Sykes' Tapered Tubes, Topham's Double-tube Boxes, Sankey's Continuous Economiser, Calvert's Continuous Economiser, Sykes' Screwed Lids, Sykes' Internal Tube Economiser, Knight & Thode's Economiser.	
ECONOMISERS HAVING CONTINUOUS CIRCULATION	71-74
Comparative tests of Circulating Economisers, by J. F. L. Crossland, M. Inst. C.E.; Michael Longridge, M. Inst. C.E.; L. E. Fletcher, M. Inst. C.E.; E. G. Hiller, M. Inst. M.E.	
CIRCULATORS AND ECONOMISERS	75-76
GREEN'S EARLY TUBULAR BOILERS	77-91
Saddle Boilers with Single and Double Tubular Legs, Egg-ended Multitubular Boilers, Vertical Tubular Boilers, Sectional Furnace Boiler, Economisers worked as Boilers, Internally-fired Tubular Boilers, Cone Tube Boiler.	
GREEN'S EARLY SUPERHEATERS	92-93
GREEN'S MODERN FUEL ECONOMISER	94-101
Quick Reversing Gear for Scrapers, Independent Engine for driving Scrapers, External Lids for ordinary pressures, Internal Lids for high pressures, Method of withdrawing Damaged Tubes, Access Lids.	
SIZE AND EFFICIENCY OF GREEN'S ECONOMISER	102-105
DIRECTIONS FOR WORKING ECONOMISER	106-108
CONCLUSION	109

Index

A

	PAGE
ACCESS LIDS FOR FLUSHING GREEN'S ECONOMISER	105
AGITATOR FOR BOTTOM BOXES, GREEN'S	16, 32
AIR BRUSH, WHITEHEAD'S	38
ALCOCK AND MASON'S ECONOMISER	54

B

BELL'S CONTINUOUS ECONOMISER	56, 60
BELL'S SPIRAL ECONOMISER	47
BLOW-OUT ARRANGEMENTS, LOWCOCK AND SYKES'	60
BOILERS, GREEN'S TUBULAR	77
BOILERS, NUMBER FITTED WITH GREEN'S ECONOMISER	3
BOTTLE ECONOMISER, ELSON'S	49
BOTTOM BOXES, GREEN'S AGITATOR	16, 32
BOXES FOR ECONOMISERS, HAWKINS'	62
BOXES FOR ECONOMISERS, LOWCOCK AND SYKES'	59, 64
BOXES, TOPHAM'S DOUBLE TUBE	67
BRUSH, LEES' STEAM	36
BRUSH, WHITEHEAD'S STEAM AND AIR	38
BURPEE'S SPIGOT JOINTS AND TAPERED TUBES	66

C

CALLIPER SCRAPER, ROBINSON'S	40
CALVERT AND TAYLOR'S ECONOMISER	49
CALVERT'S CONTINUOUS CIRCULATION ECONOMISER	68
CAPS FOR ECONOMISER TUBES (*see* LIDS)	
CHAIN SCRAPER, WHITEHEAD'S	40
CIRCULATOR, TWIBILL'S	38
CIRCULATORS *versus* ECONOMISERS	75
CONDENSER, WATTS'S ANALOGOUS TO GREEN'S ECONOMISER	2
CONDENSER, GREEN'S	80
CONTINUOUS CIRCULATION ECONOMISER, BELL'S	56, 60
CONTINUOUS CIRCULATION ECONOMISER, CALVERT'S	68
CONTINUOUS CIRCULATION ECONOMISER, CALVERT AND TAYLOR'S	49
CONTINUOUS CIRCULATION ECONOMISER, HOLT AND GALLOWAY'S	41
CONTINUOUS CIRCULATION ECONOMISER, KNIGHT AND THODE'S	70

INDEX

	PAGE
CONTINUOUS CIRCULATION ECONOMISER, LEES'	36
CONTINUOUS CIRCULATION ECONOMISER, LEES AND GARFORTH'S	52
CONTINUOUS CIRCULATION ECONOMISER, MASON AND ALCOCK'S	54
CONTINUOUS CIRCULATION ECONOMISER, SANKEY'S	68
CONTINUOUS CIRCULATION ECONOMISERS, TESTS OF	71
CRAB JAW SCRAPERS, PERKINS AND SCOTT'S	59

D

DOUBLE TUBE BOXES, TOPHAM'S	67
DOWN FEED ECONOMISER, GREEN'S	30, 33

E

ECONOMISER, BELL'S	47, 56, 60
ECONOMISER, CALVERT AND TAYLOR'S	49
ECONOMISER, CALVERT'S CONTINUOUS CIRCULATION	68
ECONOMISER, ELSON'S	49
ECONOMISER, GREEN'S FIRST EXPERIMENTAL	7
ECONOMISER, GREEN'S FIRST PATENTED	10
ECONOMISER, GREEN'S MODERN	94
ECONOMISER, HOLT AND GALLOWAY'S CONTINUOUS CIRCULATION	41
ECONOMISER, KNIGHT AND THODE'S	70
ECONOMISER, LEES AND GARFORTH'S	52
ECONOMISER, LEES' CONTINUOUS CIRCULATION	36
ECONOMISER MADE OF SEPARATE TUBES, GREEN'S	31
ECONOMISER, MASON AND ALCOCK'S	54
ECONOMISER, ORACH'S	45
ECONOMISER, PRESTWICH AND PIMBLEY'S	56
ECONOMISER, SANKEY'S	68
ECONOMISER, SYKES' GROUPED TUBE	57
ECONOMISER, SYKES' INTERNAL TUBE	69
ECONOMISER, TOPHAM'S	67
ECONOMISER, TWIBILL'S CIRCULAR	54
ECONOMISER, TWIBILL'S INTERNAL TUBE	65
ECONOMISER WITH FEED DOWNFLOW, GREEN'S	30, 33
ECONOMY EFFECTED BY GREEN'S ECONOMISERS	6
ELSON'S BOTTLE ECONOMISER	49
ELSON'S SPIRAL SCRAPER	51
EXPERIMENTAL ECONOMISER, GREEN'S FIRST CONSTRUCTION	7
EXPLOSION OF CONTINUOUS CIRCULATION ECONOMISER	42

F

FEED DOWNFLOW	30, 33
FLUTED ECONOMISER TUBES	81
FUEL, SAVING EFFECTED BY GREEN'S ECONOMISERS	6

INDEX

G

	PAGE
GALLOWAY'S CONTINUOUS CIRCULATION ECONOMISER	41
GARFORTH AND LEES' CONTINUOUS CIRCULATION ECONOMISER	52
GARFORTH AND LEES' SCRAPERS	51
GREEN, BIOGRAPHICAL SKETCH	4
GREEN'S AGITATOR FOR BOTTOM BOXES	32
GREEN'S ECONOMISER, DIRECTIONS FOR WORKING	106
GREEN'S ECONOMISER, ECONOMY EFFECTED BY	6
GREEN'S ECONOMISER INTERNAL BLOCKS AND TUBES	16
GREEN'S ECONOMISER JUBILEE	2
GREEN'S ECONOMISER, METHOD OF FLUSHING	104
GREEN'S ECONOMISER, NOTES ON SIZE AND EFFICIENCY OF	102
GREEN'S ECONOMISER, NUMBER OF BOILERS FITTED WITH	3
GREEN'S ECONOMISER WITH FEED DOWNFLOW	30, 33
GREEN'S EXTERNAL LID FOR ORDINARY PRESSURES	99
GREEN'S FIRST EXPERIMENTAL ECONOMISER	7
GREEN'S FIRST PATENTED ECONOMISER	10
GREEN'S FIRST SCRAPERS	15
GREEN'S FIRST SECTIONAL SCRAPER	17
GREEN'S INDEPENDENT ENGINE FOR DRIVING SCRAPERS	98
GREEN'S INTERNAL LID FOR HIGH PRESSURES	100
GREEN'S LID WITH COPPER-JOINT RING	29
GREEN'S LID WITH DRAW SCREW	26
GREEN'S LID WITH SOFT METAL COLLAR FOR BOLT	26
GREEN'S LID WITH TRIANGULAR BOLTS	28
GREEN'S LIDS FOR ECONOMISER TUBES	25
GREEN'S METHOD OF DRAWING AND PLUGGING DAMAGED TUBES	101
GREEN'S MODERN ECONOMISER	94
GREEN'S OVAL AND INTERNAL LIDS	28
GREEN'S OVAL LID WITH CONE JOINT FACE	29
GREEN'S QUICK REVERSING MOTION	97
GREEN'S REVERSING GEAR FOR SCRAPERS	22
GREEN'S SCRAPER WITH BALL HINGE	18
GREEN'S SCRAPER WITH DIAGONAL TEETH	20
GREEN'S SECTIONAL INTERLOCKING SCRAPER	21
GREEN'S SEPARATE TUBE ECONOMISER	31
GREEN'S EARLY TUBULAR BOILERS	77
GREEN'S EARLY SUPERHEATERS	92
GUNPOWDER USED TO REMOVE SOOT FROM ECONOMISER TUBES	9

H

HAWKINS' TOP AND BOTTOM BOXES	62
HOLT'S CONTINUOUS CIRCULATION ECONOMISER	41

INDEX

I

	PAGE
INCLINED TUBE ECONOMISER, BELL'S	56
INSTRUCTIONS FOR WORKING ECONOMISER	106
INTERNAL BLOCKS AND TUBES IN GREEN'S ECONOMISER	16
INTERNAL TUBE ECONOMISER, SYKES'	69
INTERNAL TUBE ECONOMISER, TWIBILL'S	65

J

JOINT FOR ECONOMISER TUBES, OBACH'S	45
JUBILEE OF INVENTION OF ECONOMISER	2

K

KNIGHT AND THODE'S CIRCULATION ECONOMISER	70

L

LEES' STEAM BRUSH	36
LEES AND GARFORTH'S CONTINUOUS CIRCULATION ECONOMISER	52
LEES AND GARFORTH'S SCRAPERS	51
LID, SYKES' ECONOMISER	66
LID WITH COPPER-JOINT RING	29
LID WITH DRAW SCREW, GREEN'S	26
LID WITH SOFT COLLAR FOR BOLT, GREEN'S	26
LID WITH TRIANGULAR BOLT, GREEN'S	28
LIDS FOR ECONOMISER TUBES, GREEN'S	25
LIDS FOR ECONOMISERS, SYKES' SCREWED	69
LIDS FOR HIGH PRESSURES, GREEN'S INTERNAL	100
LIDS FOR ORDINARY PRESSURES, GREEN'S EXTERNAL	99
LIDS, GREEN'S OVAL AND INTERNAL	28
LIDS, OVAL, WITH CONE JOINT FACE, GREEN'S	29
LIDS, PERKINS AND SCOTT'S	59
LOWCOCK AND SYKES' BLOW-OUT ARRANGEMENTS	59
LOWCOCK AND SYKES' BOXES FOR ECONOMISERS	59, 64
LOWCOCK AND TAYLOR'S SCRAPERS	58

M

MANNOCK'S SCRAPER	40
MASON AND ALCOCK'S ECONOMISER	54
MUNDY'S SCRAPERS	48

N

NEEDHAM'S INCLINED SCRAPERS	42
NEEDHAM'S REVERSING MOTION FOR SCRAPERS	43
NEEDHAM'S SCREW SHAFT	43
NUMBER OF ECONOMISERS MADE BY MESSRS. GREEN	3

INDEX xi.

O

	PAGE
OBACH'S TUBE JOINTS	45
OBACH'S REVERSING MOTION FOR SCRAPERS	46
OBACH'S SCRAPERS	45

P

PERKINS AND SCOTT'S LIDS	59
PERKINS AND SCOTT'S SCRAPERS	59
PRESTWICH AND PIMBLEY'S ECONOMISER	56

R

REVERSING GEAR FOR SCRAPERS, GREEN'S	22
REVERSING MOTION FOR SCRAPERS, NEEDHAM'S	43
REVERSING MOTION, GREEN'S QUICK	97
ROBINSON'S SCRAPER	40

S

SANKEY'S ECONOMISER	68
SCOTT AND PERKINS' SCRAPERS	59
SCRAPER, BELL'S	47
SCRAPER, ELSON'S SPIRAL	51
SCRAPER, GREEN'S FIRST SECTIONAL	17
SCRAPER, GREEN'S SECTIONAL INTERLOCKING	21
SCRAPER, MANNOCK'S	40
SCRAPER, MUNDY'S	48
SCRAPER, OBACH'S	45
SCRAPER, OBACH'S REVERSING MOTION FOR	46
SCRAPER, PERKINS AND SCOTT'S CRAB JAW	59
SCRAPER, ROBINSON'S	40
SCRAPER WITH BALL HINGE, GREEN'S	18
SCRAPER WITH DIAGONAL TEETH, GREEN'S	20
SCRAPERS, GREEN'S FIRST	15
SCRAPERS, GREEN'S INDEPENDENT DRIVING ENGINE FOR	98
SCRAPERS, GREEN'S QUICK REVERSING MOTION FOR	97
SCRAPERS, GREEN'S REVERSING GEAR FOR	22
SCRAPERS, LEES AND GARFORTH'S	51
SCRAPERS, LOWCOCK AND TAYLOR'S	58
SCRAPERS, NEEDHAM'S INCLINED	42
SCRAPERS, NEEDHAM'S REVERSING MOTION FOR	43
SCRAPERS, PERKINS AND SCOTT'S	59

INDEX

	PAGE
SCRAPERS, WHITEHEAD'S CHAIN	40
SCREW SHAFT, NEEDHAM'S	43
SEPARATE CONDENSER, WATTS', ANALOGOUS TO GREEN'S ECONOMISER	2
SPIGOT JOINTS, BURPEE'S	66
SPIRAL ECONOMISER, BELL'S	47
STAGGERED TUBES, BELL'S ECONOMISER	60
STAGGERED TUBES, TOPHAM'S	67
STAGGERED ROWS OF TUBES USED IN GREEN'S FIRST ECONOMISER	12
STEAM BRUSH, LEES'	36
STEAM BRUSH, WHITEHEAD'S	38
SUPERHEATERS, GREEN'S	92
SYKES AND LOWCOCK'S BLOW-OUT ARRANGEMENTS	59
SYKES AND LOWCOCK'S BOXES FOR ECONOMISERS	59, 64
SYKES' ECONOMISER LID	66
SYKES' GROUP ECONOMISER	57
SYKES' INTERNAL TUBE ECONOMISER	69
SYKES' SCREWED LIDS	69
SYKES' TAPERED ECONOMISER TUBES	67

T

TAYLOR AND CALVERT'S ECONOMISER	49
TAYLOR AND LOWCOCK'S SCRAPERS	58
TAPER ECONOMISER TUBE, BURPEE'S	66
TAPER ECONOMISER TUBE, SYKES'	67
TESTS OF CONTINUOUS CIRCULATION ECONOMISERS	71
THODE AND KNIGHT'S CIRCULATION ECONOMISER	70
TOPHAM'S DOUBLE TUBE BOXES	67
TUBES, GREEN'S METHOD OF WITHDRAWING DAMAGED	101
TUBE JOINT, OBACH'S	45
TUBES, TAPERED, BURPEE	66
TUBES, TAPERED, SYKES'	67
TUBULAR BOILERS, GREEN'S	77
TWIBILL'S CIRCULAR ECONOMISER	54
TWIBILL'S CIRCULATOR	38
TWIBILL'S INTERNAL TUBE ECONOMISER	65

W

WATER-TUBE BOILERS, GREEN'S	77
WATTS' SEPARATE CONDENSER; ANALOGOUS TO GREEN'S ECONOMISER	2
WHITEHEAD'S STEAM AND AIR BRUSH	38
WHITEHEAD'S CHAIN SCRAPER	40
WORKING OF ECONOMISER, INSTRUCTIONS FOR	106

List of Illustrations

		PAGE
FIG. 1.	LONGITUDINAL SECTION OF FIRST EXPERIMENTAL ECONOMISER	8
FIG. 2.	PLAN OF FIRST EXPERIMENTAL ECONOMISER	8
FIG. 3.	LONGITUDINAL SECTION OF GREEN'S FIRST PATENTED ECONOMISER	11
FIG. 4.	PLAN OF GREEN'S FIRST PATENTED ECONOMISER	12
FIG. 5.	ELEVATION OF SCRAPERS IN FIRST ECONOMISER	13
FIG. 6.	PLAN OF SCRAPERS IN FIRST ECONOMISER	14
FIG. 7.	FIRST SECTIONAL SCRAPER	17
FIG. 8.	SECTIONAL SCRAPERS WITH BALL HINGE	18
FIG. 9.	PLAN OF SCRAPER	19
FIG. 10.	SECTION OF SCRAPER	19
FIG. 11.	SCRAPER WITH DIAGONAL CUTTING TEETH	20
FIG. 12.	SECTIONAL INTERLOCKING SCRAPER	21
FIG. 13.	ELEVATION OF SCRAPER REVERSING GEAR	22
FIG. 14.	PLAN OF SCRAPER REVERSING GEAR	22
FIG. 15.	END VIEW OF SCRAPER REVERSING GEAR	23
FIG. 16.	END VIEW OF QUICK REVERSING GEAR	24
FIG. 17.	PLAN OF QUICK REVERSING GEAR	24
FIG. 18.	INTERNAL CONICAL LID	26
FIGS. 19 & 20.	SOFT METAL COLLAR FOR SEALING BOLT HOLE OF LID	27
FIG. 21.	TRIANGULAR BOLT FOR ECONOMISER LID	28
FIG. 22.	OVAL INTERNAL LID	28
FIG. 23.	INTERNAL LID WITH COPPER JOINT RING	29
FIG. 24.	INTERNAL OVAL LID WITH CONICAL JOINT FACE	29
FIG. 25.	ECONOMISER WITH FEED DOWN FLOW	30
FIG. 26.	ECONOMISER BUILT UP OF SEPARATE TUBES	31
FIG. 27.	BOTTOM BOX WITH MECHANICAL AGITATOR	32
FIG. 28.	ECONOMISER WITH FEED DOWNFLOW	33
FIG. 29.	LEES' REVOLVING STEAM BRUSH	36
FIG. 30.	SECTION OF TWIBILL'S SUPPLEMENTARY HEATER	38
FIG. 31.	PLAN OF TWIBILL'S SUPPLEMENTARY HEATER	39
FIG. 32.	MANNOCK'S SCRAPER	40
FIG. 33.	ROBINSON'S SCRAPER	40
FIG. 34.	HOLT AND GALLOWAY'S CONTINUOUS ECONOMISER	41
FIG. 35.	NEEDHAM'S SCREW SHAFT	43
FIG. 36.	NEEDHAM'S REVERSING MOTION FOR SCRAPERS	44
FIG. 37.	OBACH'S TUBE JOINT	45
FIG. 38.	OBACH'S SCRAPER	46

LIST OF ILLUSTRATIONS

		PAGE
FIG. 39.	OBACH'S REVERSING GEAR	46
FIG. 40.	BELL'S SPIRAL ECONOMISER	47
FIG. 41.	CALVERT AND TAYLOR'S ECONOMISER	48
FIG. 42.	PLAN OF CALVERT AND TAYLOR'S ECONOMISER	49
FIG. 43.	ELSON'S BOTTLE ECONOMISER	50
FIG. 44.	ELSON'S SPIRAL SCRAPER	51
FIG. 45.	LEES AND GARFORTH'S CONTINUOUS CIRCULATION ECONOMISER	52
FIGS. 46 & 47.	TWIBILL'S CIRCULAR ECONOMISER	53
FIG. 48.	MASON AND ALCOCK'S CONTINUOUS CIRCULATION ECONOMISER	54
FIGS. 49 & 50.	PRESTWICH AND PIMBLEY'S SPLIT TUBE ECONOMISER	55
FIG. 51.	BELL'S CONTINUOUS CIRCULATION ECONOMISER	57
FIG. 52.	SYKES' GROUPED TUBE ECONOMISER	58
FIG. 53.	PERKINS AND SCOTT'S CRAB JAW SCRAPERS	59
FIGS. 54 & 55.	LOWCOCK AND SYKES' BLOW-OUT ARRANGEMENTS	60
FIGS. 56 & 57.	BELL'S CONTINUOUS CIRCULATION ECONOMISER	61
FIGS. 58 & 59.	HAWKINS' TOP AND BOTTOM BOXES	62
FIG. 60.	LOWCOCK AND SYKES' TOP AND BOTTOM BOXES	64
FIG. 61.	TWIBILL'S INTERNAL CIRCULATING TUBE ECONOMISER	65
FIG. 62.	SYKES' ECONOMISER LID	66
FIG. 63.	TOPHAM'S DOUBLE-TUBE BOXES	67
FIG. 64.	CALVERT'S ECONOMISER	72
FIG. 65.	BLOW-OUT ARRANGEMENT OF CALVERT'S ECONOMISER	72
FIGS. 66 & 67.	GREEN'S TUBULAR SADDLE BOILERS	77, 78
FIGS. 68 & 69.	GREEN'S MULTITUBULAR EGG-ENDED BOILER	79
FIG. 70.	GREEN'S DUPLEX-TUBULAR EGG-ENDED BOILER	80
FIGS. 71 & 72.	GREEN'S SECTIONAL BOILER WITH EXTERNAL FURNACE	81
FIG. 73.	GREEN'S SECTIONAL FURNACE	82
FIGS. 74 & 75.	GREEN'S SECTIONAL FURNACE BOILER	83, 84
FIG. 76.	ECONOMISER FITTED WITH STEAM DRUM	85
FIG. 77.	GREEN'S VERTICAL TUBULAR BOILER	86
FIGS. 78 & 79.	GREEN'S TUBULAR BOILER FITTED WITH SECTIONAL FURNACE	87, 89
FIG. 80.	GREEN'S INTERNALLY FIRED TUBULAR BOILER	88
FIG. 81.	GREEN'S CONE TUBE BOILER	90
FIG. 82.	GREEN'S EARLY SUPERHEATER	92
FIGS. 83, 84, & 85.	GREEN'S MODERN ECONOMISER	94, 95
FIG. 86.	GREEN'S REVERSING GEAR FOR SCRAPERS	97
FIG. 87.	INDEPENDENT ENGINE FOR DRIVING SCRAPER	98
FIG. 88.	GREEN'S EXTERNAL LID FOR ORDINARY PRESSURES	99
FIG. 89.	GREEN'S INTERNAL LID FOR HIGH PRESSURES	100
FIG. 90.	METHOD OF DRAWING DAMAGED TUBE AND INSERTING STOP FERRULES	101
FIG. 91.	METHOD OF FLUSHING BOTTOM BOXES	104
FIG. 92.	PATENT ACCESS PIPE FOR FLUSHING ECONOMISER	105

List of Plates

	FACING PAGE
GREEN'S MODERN ECONOMISER, ETC.	TITLE
CASTING TUBES VERTICALLY	7
A CORNER OF THE FOUNDRY	35
A CORNER OF THE FITTING SHOP	57
SCRAPER GEAR FITTING ROOM	71
HYDRAULIC PRESSES	109

HISTORY OF THE ECONOMISER

INTRODUCTION

THERE is probably no adjunct in connection with the working of Stationary Steam Boilers whose economical value is more generally recognised than that known as the ECONOMISER, or FEED-WATER HEATER; and with the invention and development of which the name of GREEN has become so largely identified. Indeed it is no exaggeration to say that wherever Stationary Boilers of any size or number are at work, there also will be found a "Fuel Economiser."

The apparatus consists in the main of a stack of tubes, arranged vertically in the flue leading from the boiler to the chimney, and designed to utilize the waste heat in the gases passing off from the furnace. This is accomplished by absorbing the low temperature heat of the gases in heating the feed-water, which is pumped through the Economiser in the first instance, before its entry into the boiler.

It is this fundamental principle of heating the feed-water in a Separate Vessel quite apart from the boiler, and thereby utilizing the heat in the waste gases passing to the chimney, which constitutes the distinctive feature of the first Mr. Edward Green's invention—an invention which in its influence on boiler economy and design is analagous, in many respects, to Watts' application of the Separate Condenser to the Steam Engine.

It was in 1845 that the first Mr. Edward Green made his earliest experiments with the working of this apparatus, which in the present year enters upon its jubilee, and which is now regarded as the almost indispensable adjunct of a steam power plant by whose aid the surplus heat, escaping from the boilers, is arrested and utilised.

The reputation of the apparatus is not confined to Great Britain alone. It is in operation all over the Continent, and is as well known in the spinning mills of Russia and India as those of Lancashire and Yorkshire. Throughout the large manufacturing districts of the United States it is almost universally adopted, and is as equally noted among the gold mines of South Africa as in the textile factories of China and Japan; while its credit as a saver of fuel has been so well established and its fame so widely spread, that there is hardly a single electrical installation in this country or the United States for lighting or traction that does not include a Green's Economiser. Some measure of the general recognition of the value of the Economiser is afforded by the fact that since Messrs. Green commenced business they have supplied the apparatus to over 150,000 boilers, representing some thirty millions indicated horse-power.

Having regard to the importance of the primary invention and of the far-reaching consequences that have followed its development, a brief outline of its first introduction and subsequent growth during the course of fifty years will, it is thought, be of interest to that large section of engineers who are especially associated with the design and working of steam power installations.

A chapter has been devoted to a brief account of some early forms of tubular boilers constructed by Messrs. Green. This has been introduced rather as a matter of curiosity, in view of the interest that has recently been excited in steam generators of this type, and as showing some phases of the early development of this class of boiler. Messrs. Green, however, are not now engaged in the construction of tubular boilers, having given up this section of their business some 20 years ago in order to devote themselves solely and exclusively to the development and manufacture of the Economiser.

THE
Inventor of the Economiser

IN tracing the progress of the Economiser from the original invention to its present development the reader of the following pages will not fail to be struck with the versatility and readiness of resource displayed by the inventor in overcoming the practical difficulties that more or less attend the development of an engineering apparatus of this kind. As the late Mr. Edward Green was, in many respects, a remarkable man, the following brief biographical sketch of his career may not be out of place.

The subject of our remarks was born on the 6th of January 1799, at Wakefield, Yorkshire. His faculty for invention and mechanical pursuits was manifested at an early age, and without going into details it may be stated that it largely determined his choice of a career, and he became apprenticed in due course to a firm of engineers in the city referred to. His daily work afforded opportunities for his inventive talents, of which he was not slow to take advantage, and his ideas manifested themselves in a variety of directions ranging from steam engines and agricultural machinery to various domestic appliances. His genius for invention was irresistible, and he appeared never satisfied unless engaged in designing some original device or improving some existing arrangement.

Before Mr. Green began to devote his whole time and energies to the apparatus that bears his name, he was engaged for many years in general mechanical work, and in the construction of large pumping engines for the deep coal mines of Yorkshire, as well as others for draining the Fens in Lincolnshire and Norfolk. As engineering work this may perhaps not be thought much of at the present day, but it is to be remembered that at the beginning of the present century the science of steam engineering was only in its infancy, and the laws of heat but imperfectly understood, while the available data as to the strength both of materials and structures were of a very meagre character. The work of the engineer was thus fenced round with great difficulty, and it was frequently necessary for him to traverse untrodden ground and rely largely on his individual experience and judgment.

In addition to his mechanical and engineering pursuits Mr. Green displayed a considerable amount of interest and public spirit in connection with a scheme for the supply of water for manufacturing purposes to his native town, while the subject of sanitary reform possessed in him a strong and ardent advocate at a time when sanitary reformers were few, and the importance of the subject was but imperfectly realised. One of his suggestions in connection with this branch of engineering was that the main sewers of large towns should be connected to a common tall ventilating shaft, having a furnace at its base, through which the noxious vapours could be passed in order to render them innocuous before being discharged into the atmosphere. Soon after the introduction of gas lighting he instituted a method for automatically measuring and regulating the supply of gas to the public lamps.

Though the patent for the "Fuel Economiser" was taken out in 1845. it was not until after the great Exhibition of 1851 that steam users seriously recognised the great opportunity for saving that was, by its means, placed within their reach. From that date, however, its success was assured, and the large amounts of coal that were economised by those steam users who gave it a trial, forced it upon the attention of the remainder of the manufacturing community, with the result that the advantages of the Economiser became more and more widely known, and at length universally recognised. At the present day the invention is used by boiler owners all over the globe, and it is computed effects a saving of upwards of one-and-a-quarter millions sterling per annum.

Mr. Edward Green died at Wakefield in the year 1865, and was succeeded in the business he established by his son, Mr., now Sir Edward Green, Bart., the present head of the firm, who was joined by his second son, Mr. Frank Green, in 1883.

CASTING TUBES VERTICALLY.

THE
First Experimental Economiser

THE first Economiser that was made by Mr. Edward Green, and which formed the experimental basis of his subsequent designs, is shown in the illustrations on page 8. It consisted of a circular group of 30 cast-iron tubes, 4ins. in diameter by 9ft. in length, connected at the top and bottom to hemispherical chambers, the cold feed being introduced into the lower one, and after rising through the tubes passing off through a branch at the top to the boiler. At the bottom of the lower hemispherical chamber a blow-off tap was fixed, while a safety valve was mounted on the upper one. The vessel was practically an independent boiler, except that it was entirely filled with water, while the ratio of heating surface to cubic capacity was much larger and the range of temperature much less than in an ordinary steam boiler. These primary distinctive features and dimensions are, it is interesting to note, maintained even in the most highly-developed form of the apparatus in use at the present day.

The first experimental trials that were made with this apparatus led great expectations to be formed of its success. The water on its passage through the tubes was raised to a considerable temperature, while steam was formed in such quantities that it was frequently blown-off from the safety valve; more important still, a large saving was found to be effected in the amount of coal required. The expectations thus naturally raised were not, however, permanently maintained. In a few weeks time it was observed that the escape of steam from the safety valve became less frequent, the temperature of the water passing off from the Economiser gradually lower, while the economy in coal correspondingly dwindled, in addition to which the draught became seriously impaired.

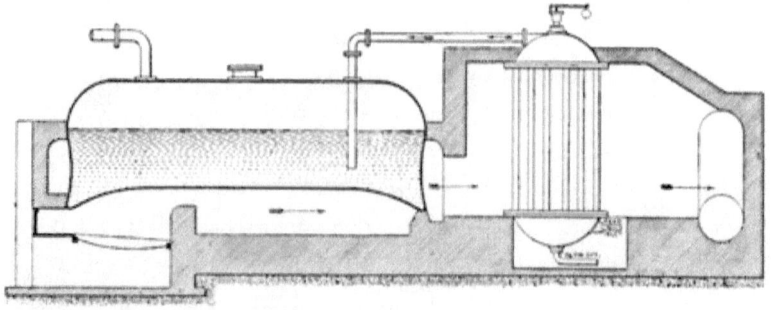

FIG. 1. LONGITUDINAL SECTION OF FIRST EXPERIMENTAL ECONOMISER.

This gradual falling off in efficiency naturally caused some little disappointment and apprehension as to the value of the apparatus. Mr. Green, however, was convinced that there must be some reason for the change, and proceeded to investigate the cause. With this object the brickwork surrounding the apparatus was taken down, when the source of the trouble was at once revealed.

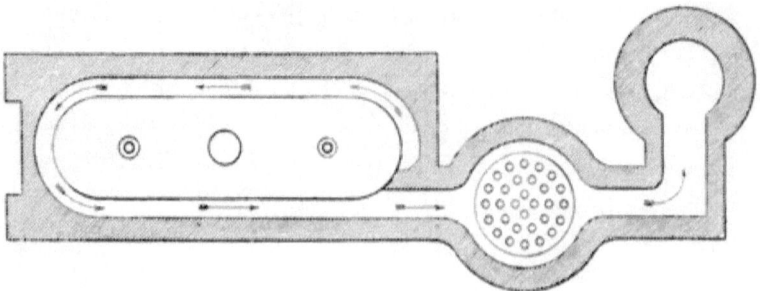

FIG. 2. PLAN OF FIRST EXPERIMENTAL ECONOMISER.

The stack of tubes was choked with soot, which not only effectually screened them from the action of the heat, but also seriously blocked up the flue, and thus accounted for the defective draught. The tubes were cleaned, the soot removed from the flue and the apparatus re-started, when the good results first noted were again obtained. The water was raised to a high temperature, steam blew off at the safety valve, and a large economy was effected in the coal required for the boiler. As before, however, these good results were only maintained for a short time, and were followed by a gradual lowering of efficiency as in the first test. The brickwork was again taken down, and the apparatus thoroughly cleaned. Flue doors were then fitted to the chamber, through which shavings were inserted and fired at intervals in order to burn off the soot which accumulated on the tubes, its removal being further facilitated by the explosion at intervals of small charges of gunpowder.

These devices proved partially successful, but they could scarcely be considered satisfactory or of a character to commend themselves to practical men. This led Mr. Green to design some automatic method of preventing the accumulation of soot. After a little scheming he succeeded in effectively doing this by means of mechanical scrapers, and there can be little doubt that the subsequent success of the Economiser as a practical apparatus was largely due to the mechanical method of removing the soot first planned by him. The idea it may be here stated, was solely and wholly his, and his claims to the originality of this or indeed the whole invention have never been disputed. After a series of trials with this first experimental apparatus Mr. Green proceeded to protect his several designs by means of letters patent, which were granted to him on December 10th 1845. Patent No. 10,986.

The Development
of
Green's Economiser

The First Economiser Patent

AS Patent No. 10,986, in the year 1845, is practically the foundation upon which all subsequent Economiser improvements are more or less based, it may be of interest to reproduce the statement of claim, as well as the drawing by which it was illustrated, showing the application of the Economiser to a range of wagon boilers. The following is a copy of the specification:

E. Green
Patent No. 10,986,
1845

"I, the said Edward Green, do hereby declare that the nature of my invention and improvements, and the manner in which the same is to be performed, is particularly described and ascertained in and by this instrument in writing. My invention consists of a new combination or arrangement for the purpose of collecting and **applying to useful purposes the residual heat of the air or gases passing from the flues of steam boilers,** or other boilers and furnaces, or of either, after such heated air or gases have in ordinary cases ceased to act with useful effect, and are permitted to make their escape and be wasted. This combination or arrangement consists, first, of an apparatus being a series of separate pipes, tubes, or chambers, hereinafter called pipes, placed vertically, connected at the ends by pipes, or by the cisterns hereinafter described, through which vertical pipes, water or other liquids are caused to rise slowly and gradually upward, while at the same time the heated air and gases passing, or having passed from the flue or flues of boilers or furnaces, are made to circulate transversely through the spaces or compartments between and around the pipes, and to remain sufficiently

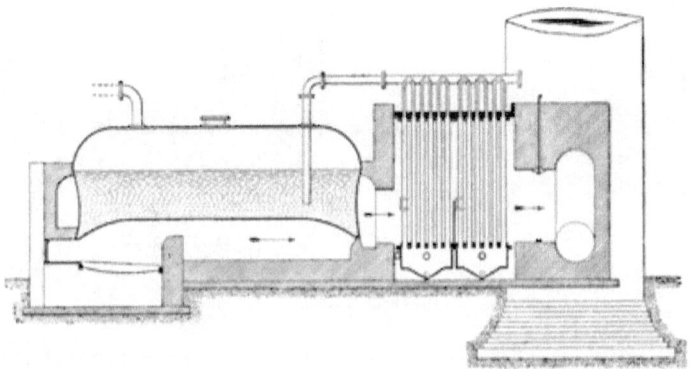

FIG. 3. LONGITUDINAL SECTION OF GREEN'S FIRST PATENTED ECONOMISER.

long in contact with them to impart to the water or other bequest contained in and passing through the pipes so much of the surplus or ordinarily wasted heat as may be required for the occasion. By this arrangement of the pipes the gravitation of the water and liquid contained therein is made to aid the conduction of the heat from the heated air and gases in contact with them, the portions of the water or liquid in contact with the pipes continually ascending as they become heated, and other and colder portions succeeding, whilst at the same time any particles of steam that may be generated during the ascent escape upward not only without impeding the general flow of the water or liquids, but even assisting the action of the upward current.

"Secondly, of an apparatus of cisterns placed at the bottom of the vertical pipes, into which the water or liquid to be heated flows and with which the whole of the pipes communicate, and of a corresponding number of cisterns at the top of the pipes, with which also they

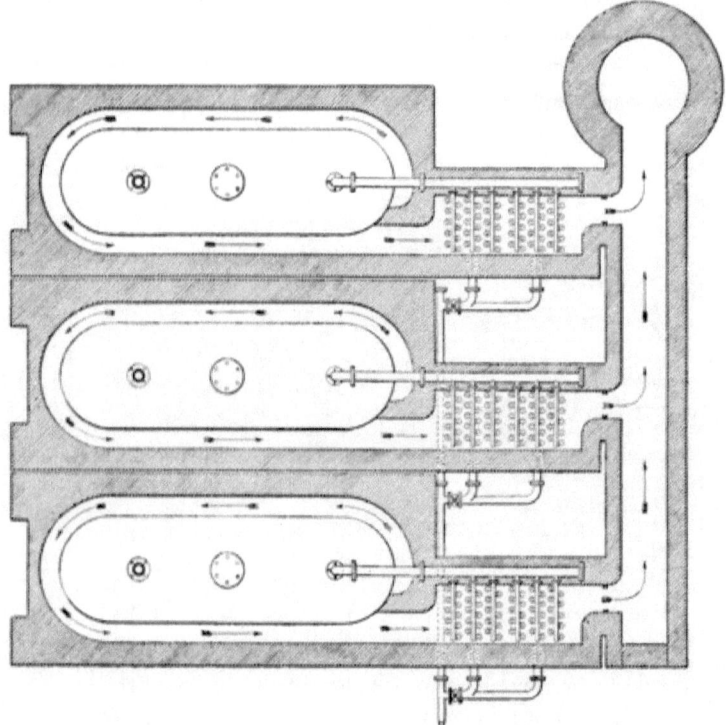

FIG. 4. PLAN OF GREEN'S FIRST PATENTED ECONOMISER.

communicate, and which upper cisterns form a reservoir of the heated water or liquid from whence it may be conveyed to feed the boilers of steam engines, or for application to any other useful purpose or manufacture.

"The bottom cisterns are so constructed as to collect any impurities or ingredients that may be mechanically suspended in the water or liquids, and to admit of their being discharged at pleasure by a cock or plug at the bottom, and the upper cisterns are furnished with movable lids for ready access to the pipes, thereby enabling the crust or deposit which necessarily forms in the inner surfaces to be cleaned off or removed when required which crust or deposit, if suffered to accumulate, would considerably impair the absorbing and conducting power of the pipes, and thus occasion the loss of much heat and corresponding waste of fuel. By these several arrangements of the pipes and of the cisterns, and the manner of passing the heated air or gases through the spaces and compartments between and around them, the hottest portion of the gases impinging on the pipes nearest the fire, and the colder on those more remote, a flow of water through the pipes is caused proportionate to the heat applied, being quicker at the hotter part and slower at the colder part of the apparatus, **thus permitting the flow of water** from the lower to the upper cisterns **through each pipe to adjust itself to the quantity of heat received**

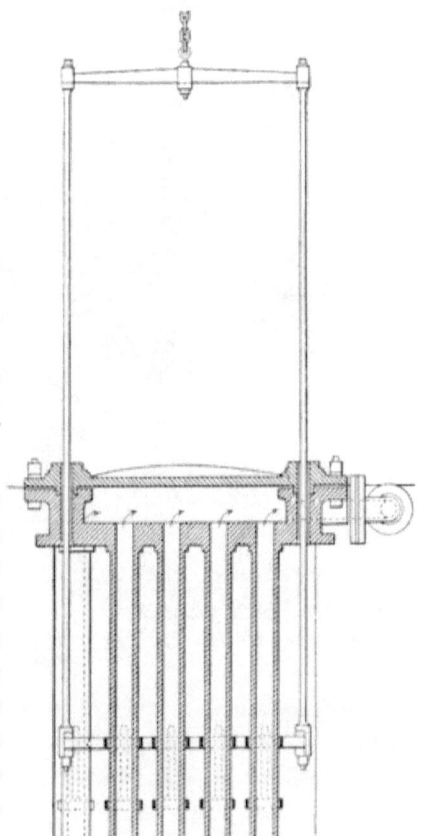

FIG. 5. ELEVATION OF SCRAPERS IN FIRST ECONOMISER.

by that particular pipe, and thereby equally to abstract the surplus heat from the heated air or gases. The arrangement has also the further advantage that the upper and hottest portions of the air or gases flowing through the spaces between and around the pipes are brought into contact with the upper and hottest portion of the water or liquid flowing through the pipes and the colder with the colder portion in the lower parts, thus abstracting the heat in the most effective manner.

"Thirdly, of an apparatus of Scrapers attached to a frame and made to encircle the pipes, which are moved upwards and downwards with a continuous and alternating motion, so as to keep the pipes continually free from any deposit of soot, thus permitting always the full action of the heated air and gases, which otherwise would be soon impaired or destroyed; these Scrapers are balanced by chains connecting them and passing over pulleys, and they may be worked either in connection with the engine, if there be one, and by a self-acting arrangement of suitable pulleys and gear made to alternate three or

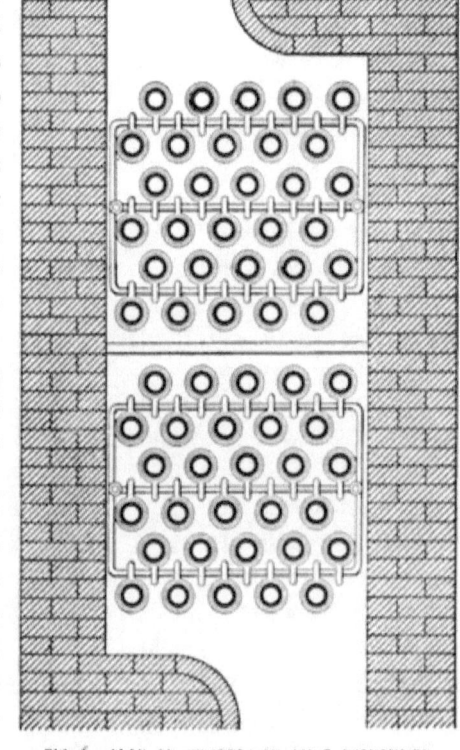

FIG. 6. PLAN OF SCRAPERS IN FIRST ECONOMISER.

four times an hour, or where no engine is attached, they may be moved by hand, as occasion may require."

The arrangement of Economiser described in this specification, and which is illustrated on pages 11-14, practically embodies the leading features of nearly all subsequent designs of this apparatus, including the movable covers in order to obtain access to the interior of the tubes ; the use of automatic scrapers ; and also the arrangement of the flow of water and flow of the gases, so as to secure as far as possible a maximum and uniform temperature difference.

On reference to the drawings it will be seen that the Economiser for each boiler consisted of 60 vertical tubes, in two sets of 30 each, connected to a large bottom cistern or box, the tubes being arranged in 12 staggered rows of five each, with a transverse connecting pipe for the hot water at the top of each pair of rows. The large cisterns at the bottom of the Economiser were designed with a view to collecting any sediment or impurity in the feed-water, which was blown off periodically as required.

The arrangement of scrapers which were used will be readily comprehended on a reference to the enlarged drawings, Figs. 5 and 6, showing this portion of the apparatus. There were, it will be observed, two frames to the set of 60 tubes, each frame containing 30 scrapers, so that one set exactly balanced the other, the two being connected by chains passing over pulleys, whereby the apparatus could receive an alternating motion by hand or machinery as desired.

The scrapers in this first arrangement consisted of two plain hoops which embraced the tubes at a little distance apart, and thus formed a circular slide. These proved fairly efficient for removing the soot when first fixed, but in the course of time, as the hoops became enlarged by wear, they were apt to lose their efficiency, and instead of scraping the tubes clean, to slide over the layers of soot and smooth them down into a thin, hard, non-conducting cake. With a view to remedy this defect, which several years' experience made manifest, the inventor next devoted his attention to improving the design of scrapers, with the results shown in pages 17 to 21.

E. Green and E. Green, Jr. Patent No. 877, 1858. This patent, taken out by Messrs. Green in 1858, appears to treat principally of agitators to prevent incrustation, and also of internal water tubes. The following were the salient points:

"This invention has reference to and consists of improvements upon three former inventions, for which letters patent were granted to Edward Green on December 10th 1845, December 10th 1853, and September 13th 1856, respectively. In the boiler or furnace flues, through which the heat from the furnace escapes, we place a series of pipes, chambers, or passages, resembling in their general features those described in the specifications of the former patents mentioned. Within these pipes we place either solid blocks of metal, earthenware, or other suitable substance or hollow pipes, which may be filled with water, leaving an annular space for the passage of steam between the outer pipes and the inner blocks or pipes .... The object of placing water-pipes within the outer pipes is the same, together with that of imparting heat to the water contained in them, which water may be used for feeding the boiler or for other purposes, or may even pass off from the apparatus in the form of steam .... And having now described the nature of the said invention, and in what manner the same is to be performed, we declare that we do not confine the use of our improved apparatuses to the heating of steam and water, as the same are intended to be employed for heating air and all other fluids, to the heating of which they may be applicable, and we claim: Firstly—placing water or other pipes, chambers, or passages in the flues of furnaces, and furnishing the same with internal blocks . . . Secondly—placing water or other pipes, chambers, or passages within steam pipes, chambers, or passages, and these again in the flues of furnaces essentially as hereinbefore described."

SCRAPERS FOR ECONOMISER TUBES

**E. Green
Patent No. 2,142,
1856**

In the specification of Patent No. 2,142, 1856, Mr. Edward Green proposed a number of various designs for scrapers, all more or less improvements on the type that had up to this date been adopted. The one, however, which proved most satisfactory in practice, and may be said to have formed the lines on which subsequent development took place, is shown in the accompanying illustrations, Fig. 7. The scraper instead of being made in a complete ring as previously, was constructed in two halves, which were hinged on to pins carried by the cross-bars of the frames. The Scrapers had sharp cutting edges, and while the tendency of the motion was to cause the scrapers to clasp the tube and thus keep it clean, they were made with a little play so as to allow them to open slightly if need be, and thus accommodate any accidental inequalities existing on the outside of the tube.

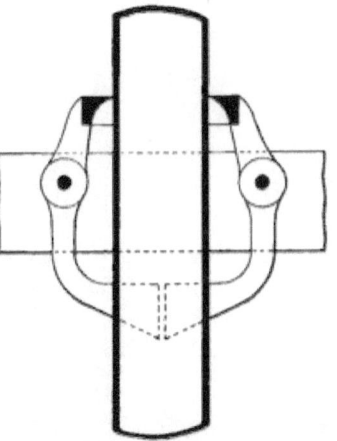

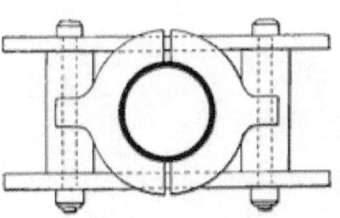

FIG. 7. FIRST SECTIONAL SCRAPER.

**E. Green
Patent No. 999,
1870**

It has been already pointed out that the efficiency of an Economiser very largely depends upon maintaining the exterior of the tubes in clean condition and free from soot or other non-conducting coating, so as to enable the hot gases passing over the outer surface of the tubes to give up their heat as rapidly and freely as possible to the

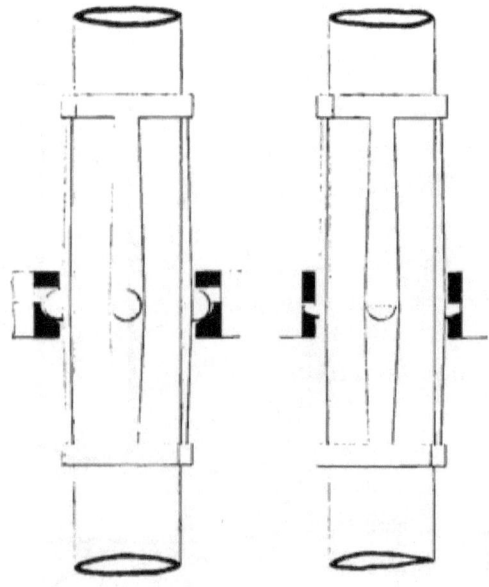

FIG. 8. SECTIONAL SCRAPERS WITH BALL HINGE.

water within them. To this end considerable attention and thought have been devoted to the scraping part of the apparatus, and it was with a view to render this more efficient for the purpose that the arrangements embodied in Patent No. 999, granted to Mr. E. Green in 1870, were designed.

On comparing these designs with those which were previously proposed it will be evident that they constituted a great improvement. The main object of the invention was to so shape the scrapers and the parts connected with them as to permit

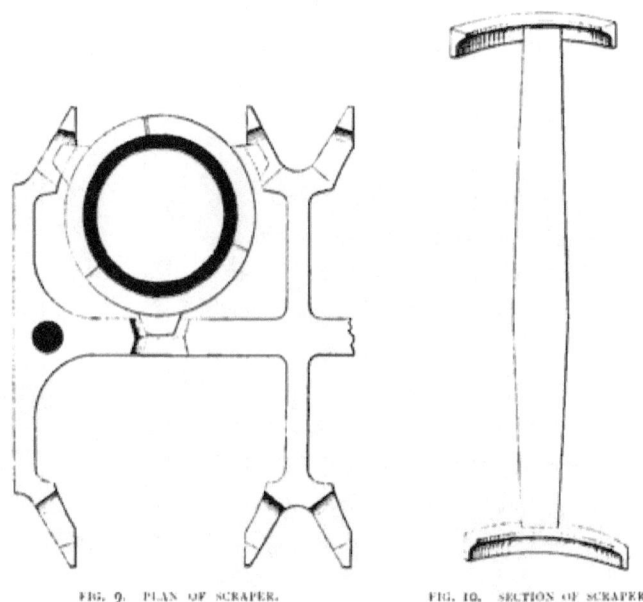

FIG. 9. PLAN OF SCRAPER.   FIG. 10. SECTION OF SCRAPER.

them to conform to the sides of the tubes under all conditions without fear of sticking or gagging, an occurrence which sometimes gave rise to trouble and annoyance. In the improvements under consideration the method proposed to overcome the difficulty was briefly to construct the scrapers in three or four sections, each section being pivotted upon an inclined surface or upon the edge of a lug or projection on the lifting bar or frame.

Each section was thus free to slide down the incline and keep its face or scraping edge against the tube. The supporting pivots being placed a little below the centre of the scraper caused it naturally to incline towards the tube, and thus clasp the surface of the tube more effectively in its upward traverse. With a view to embrace the complete circle of the tube the circumferential abutments at the top and bottom were placed a little out of line vertically so as to break joint.

The action of the scrapers and the various improvements referred to will be readily comprehended from the accompanying sketches taken from the patent specification, Fig. 10 being a detailed view of the scraper, while Fig. 8 shows elevations of two types of scraper in position on the tubes, and Fig. 9 a plan showing the disposition of the scrapers with respect to the carrying frame.

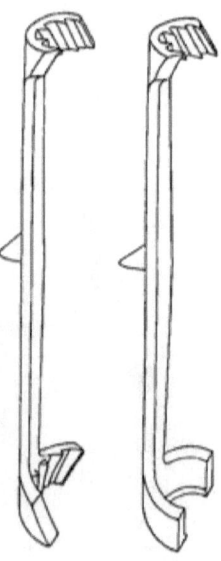

FIG. 11. SCRAPERS WITH CUTTING TEETH.

Sir E. Green Patent No. 13,804, 1886 Patent No. 13,804, 1886, by Sir E. Green, embodied some further suggestions for the improvement of the efficiency of the scrapers for Economiser tubes. The proposal consisted in arranging a number of diagonal blades or teeth in the portion of the scraper clasping the tubes, as shown in Fig. 11. A variety of designs are shown in the specification, all substantially embodying the same principle, but only one or two examples are here illustrated.

GREEN'S ECONOMISER

Sir E. Green Patent No. 23,900, 1892

Patent No. 23,900, 1892, relates to a further improvement by Sir Edward Green, respecting scrapers for Economiser tubes. These are shown in the accompanying sketch, Fig. 12. It will be observed that each circular scraper is composed of three segments, and each segment at its junction with the other two is scarphed or bevelled. This bevelling is not the same in each case, but is so arranged that the three segments can be readily dropped into the carrying frame by simply placing them in a certain order, and just as easily removed when occasion requires, although in conjunction they practically interlock each other. Each segment is fitted with a lug, having an inclined face on its underside, so that in descending the scraper slides down the tube by the pressure of its own weight. On the upward stroke, however, the wedge-action of the lugs, combined with the adjustable arrangement of the segments, causes the scraper to clasp the tube tightly, and, by means of the circular cutting edges at the top and bottom, to thoroughly scrape the surface.

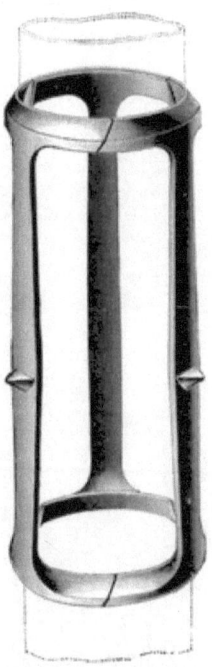

FIG. 12. SECTIONAL INTERLOCKING SCRAPER

DRIVING AND REVERSING GEAR FOR SCRAPERS

**E. Green
Patent No. 2,184,
1866**

In 1866 several important improvements were effected in the driving and reversing gear for the scrapers, and these were embodied in Patent No. 2,184, granted to Mr. E. Green. The principal features of the invention consisted in the employment of a single reversing action

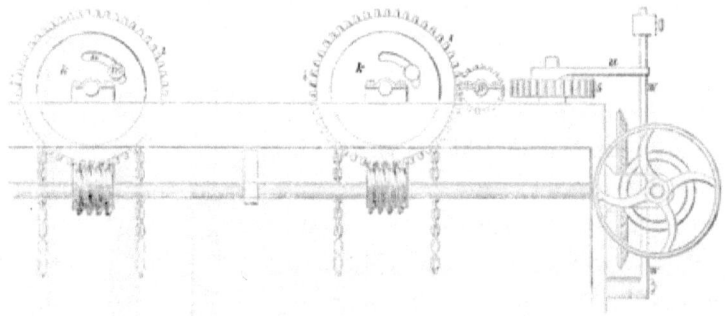

FIG. 13. ELEVATION OF SCRAPER REVERSING GEAR.

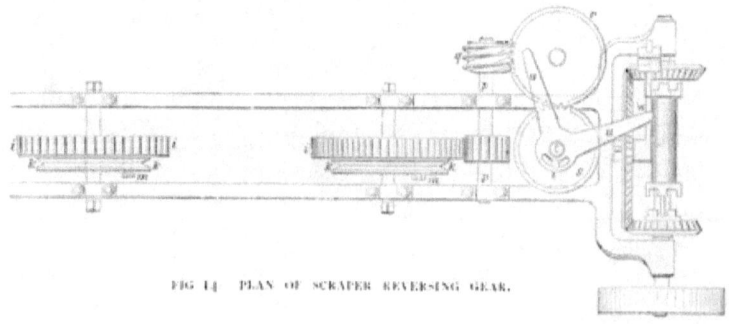

FIG. 14. PLAN OF SCRAPER REVERSING GEAR.

for several pulleys, as well as a reciprocating motion deduced from a continuous positive movement of the gearing. This combination is coupled with facilities for altering the length of traverse, and a method of securing the wheels upon the shafts and transmitting the motion so as to avoid straining the parts when the clutch box is getting into gear.

The method by which the several improvements were effected will be best realised by a reference to the accompanying illustrations. Figs. 13, 14, and 15 taken from the Patent Specification. Fig. 14 is a plan of the arrangement, **i.i.** being the worm wheels to which the chain wheels, **k.k.**, operating the scrapers, are attached. One of the wheels, **i.**, by means of the pinion and

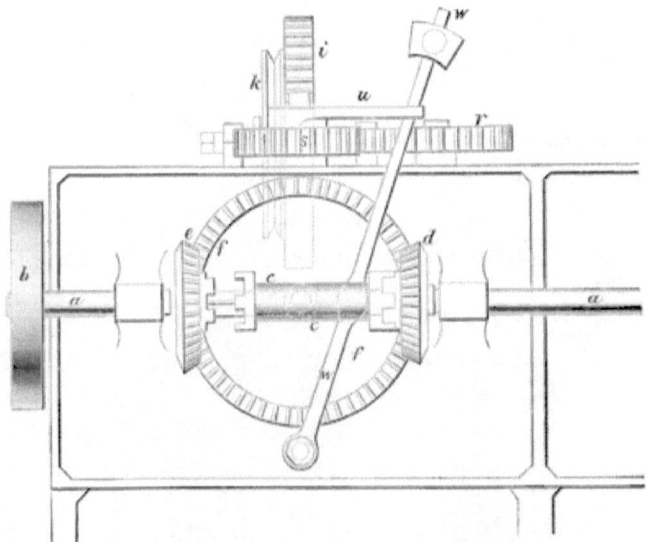

FIG. 13. END VIEW OF SCRAPER REVERSING GEAR.

worm, **p.q.**, in conjunction with the wheels, **r.** and **s.**, operates the bell crank lever, **u.**, which in its turn acts on the knock-over lever and weight, **w.**, Fig. 15. By adjusting the stop, **t.**, in the circular slot it will be evident that the motion of the scraper chains,

Fig. 13. can be reversed at any desired point, also since the chain wheels, **k.k.**, are secured to the driving wheels, **i.i.**, by means of studs, **m.**, any particular set of scrapers may be disconnected by simply slackening the stud, without interfering with the working of the others.

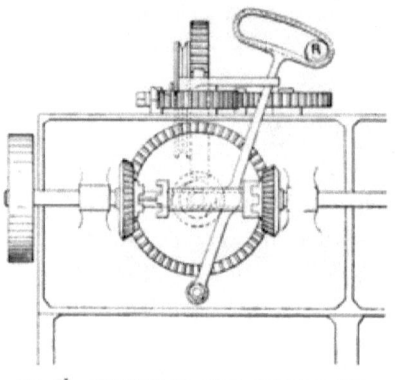

FIG. 16. END VIEW OF QUICK REVERSING GEAR.

Sir E. Green Patent No. 7,623, 1887 With a view to render the reversing motion in connection with Economiser Scrapers more quick and positive in its action, Sir E. Green, in Patent No. 7,623, 1887, proposed to supplement the action

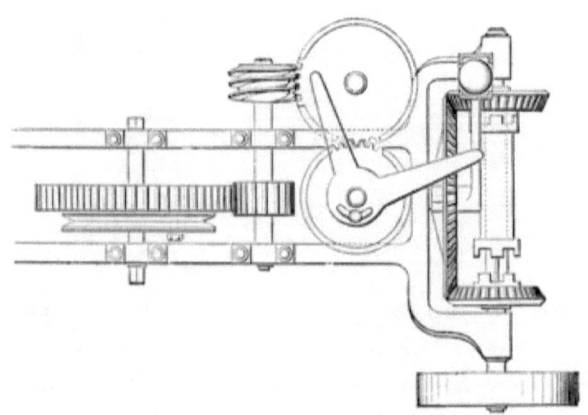

FIG. 17. PLAN OF QUICK REVERSING GEAR.

of the ordinary reversing arrangement that has already been described, by equipping the clutch lever with a weight-race. In this race a live ball, roller, or sliding weight, by rapidly changing the position of the centre of gravity of the lever when on the dead centre, secured the more prompt and effective action of the reversing mechanism. The manner in which it is proposed to apply this arrangement will be readily understood by reference to the accompanying sketches, Figs. 16 and 17, in which **R**. represents the live or sliding weight.

Sir E. Green Patent No. 23,901, 1892 — Patent No. 23,901, 1892, by Sir E. Green, has reference to improvements in the mitre gear wheels used in the reversing mechanism connected with the scrapers, the said improvement consisting in the use of shrouded wheels to prevent the rattling and shock sometimes produced when the reversing mechanism is put into operation.

LIDS AND SEATINGS FOR ECONOMISER TUBES

E. Green Patent No. 3,794, 1882 — The rise in steam pressures which followed the introduction of the compound and triple expansion engine involved, as a consequence, several modifications in the design of Economiser details. Prominent amongst these was the method of securing the lids of the sight holes in the top boxes, affording access to the interior of the tubes. Up to this date these lids were of the external type, held in position by an internal crossbar and screw. For very high pressures, however, it was desirable there should be no risk of the lids being blown off in the event of the crossbar being weakened by corrosion or straining.

With a view to obviate any risk of this kind the type of lid shown in the accompanying sketch, Fig. 18, and which is taken from the drawing accompanying Patent No. 3,794, 1882, was designed. The cover was of the internal type, and made slightly conical, with the small end of the cone looking upwards, so that the tendency of the pressure in the interior of the Economiser was to force the cover more tightly in its place. The cover, it will be noticed, was provided on its upper side with a hole for the reception of a Lewis bolt, to which a screw could be attached, and by means of the movable external cross bar or saddle the lids drawn in place. The conical joint faces were in practice made plain, and it was found this was adequate for tightness, though the patent was made to embrace a screw-thread on the conical facing of the lid, having the same pitch as the draw-screw in case this was found to be desirable.

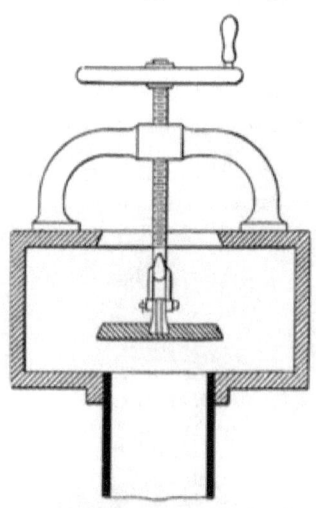

FIG. 18. INTERNAL CONICAL LID.

E. Green Patent No. 2,623, 1884 Patent No. 2,623, 1884, had reference to a method proposed by Mr. E. Green for sealing the holes in Economiser caps in cases where these caps or lids were of the external type and held in position by a central bolt and internal cross-bar. The novelty of the invention consisted in the use of a conical soft metal collar to fill the space between the bolt and the hole in the cover. This collar was expanded by the pressure of the nut in tightening up, and thus made a steam and water-tight joint. The device was not adopted to any extent, but

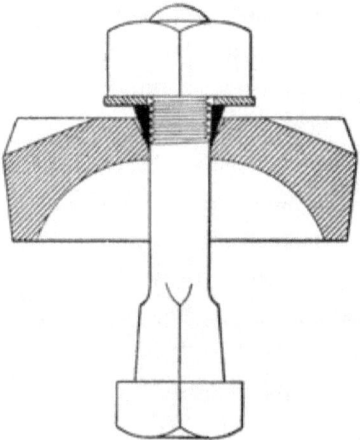

FIG. 19. SOFT METAL COLLAR BEFORE TIGHTENING BOLT.

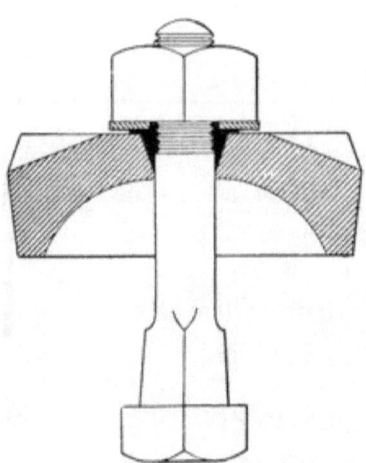

FIG. 20. SOFT METAL COLLAR AFTER TIGHTENING BOLT.

is described here in order to render the record of the various patents complete. See Figs. 19 and 20.

Sir E. Green Patent No. 3,867 1889

In the year 1889, Patent No. 3,867, Sir E. Green described a proposed improvement in the form of the bolt head securing the cap of the Economiser tube to the internal cross-bar. It was found in some cases that the head of the bolt under the cross-bar was seriously eaten away by the action of corrosive water, and the holding power of the bolt thereby seriously impaired. With a view to prevent this it was proposed to make the shank of the bolt of tapering triangular form, so as to dispense with the ordinary head, as shown in Fig. 21.

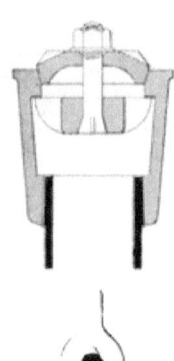

FIG. 21. TRIANGULAR BOLT FOR ECONOMISER LID.

Sir E. Green Patent No. 9,546 1892

Patent No. 9,546, 1892, relates to certain proposed improvements by Sir E. Green, in the construction of the caps or lids for the top boxes of Economisers. The design proposed is shown in Fig. 22, which gives a longitudinal section through the top box and also a plan of the lid. The lid, it will be observed, is made oval for its insertion through the hole in the box, which is fitted with a depending lip or narrow-joint surface against which the face of the cover bears.

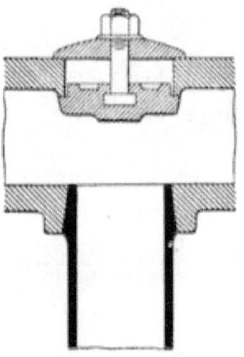
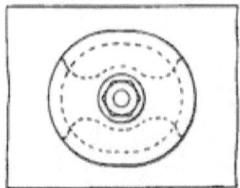

FIG. 22. OVAL INTERNAL LID.

Sir E. Green Patent No. 18,674, 1892

In 1892, another patent, No. 18,674, was granted to Sir E. Green, for two designs of conical lids. These are shown in Figs. 23 and 24. In Fig. 23 the cover is circular, with its larger diameter slightly less than the smaller diameter of the conical hole in the box so as to permit of its insertion, the joint being made by means of a

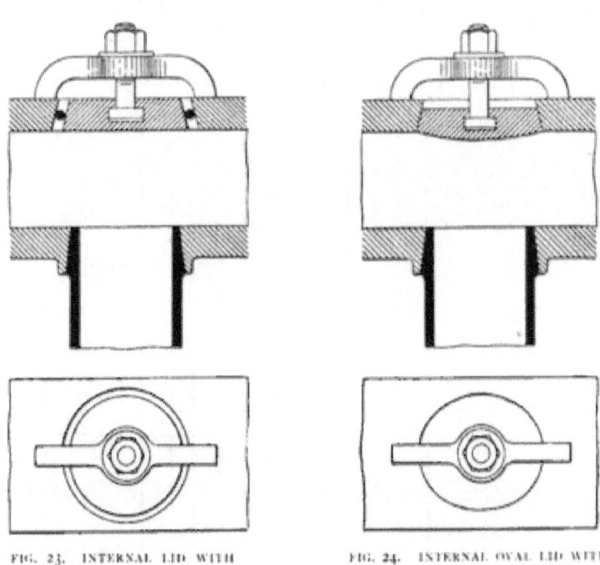

FIG. 23. INTERNAL LID WITH COPPER JOINT RING.

FIG. 24. INTERNAL OVAL LID WITH CONICAL JOINT FACE.

copper or other soft metal ring of circular section, which would permit of being sprung through the hole and placed in position over the conical lid. This could then be tightened up, the pressure on the packing ring when the lid was drawn up into place making the joint. In the design shown in Fig. 24 the lid is oval, so that it can be made of the same size as the hole in the top box, and the joint formed metal to metal without the aid of a packing ring, as shown in the previous design.

SECTIONAL TOP AND BOTTOM BOXES

**E. Green
Patent No. 8,178,
1885**

Patent No. 8,178, 1885, granted to Mr. E. Green, had reference to a method of arranging the Economiser tubes in such a way that one or more of the tubes were separated from the others by a partition or

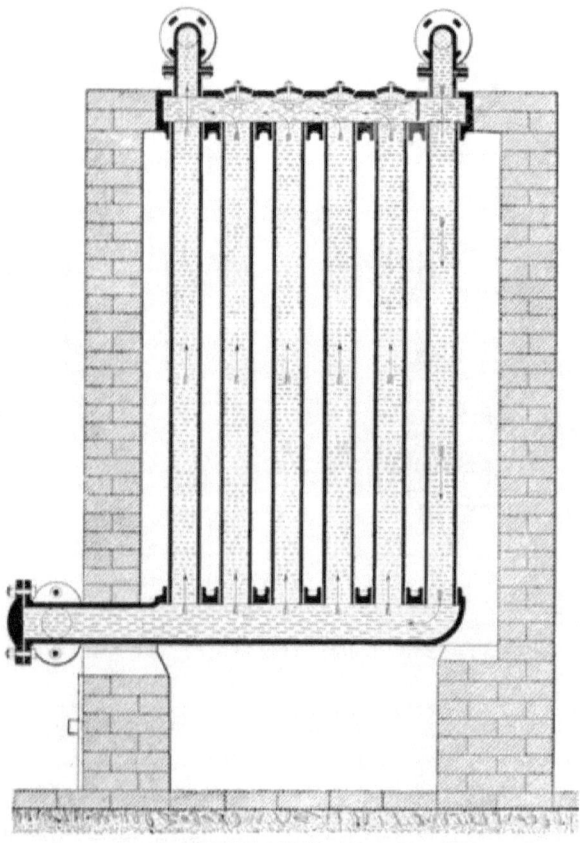

FIG. 25. ECONOMISER WITH FEED DOWN FLOW.

diaphragm in the upper box, so that the cold feed which entered from the main situated at the top of the Economiser was made to pass in a downward stream on one side of the partition into the bottom boxes, from whence it ascended in a series of upward currents into the top boxes. Another portion of the Patent had reference to the construction of the Economiser, so that each tube was a separate casting and interchangeable, and thus the Economiser could be built up, if desired, of any required dimensions. The features of these proposed modifications of design are shown in Figs. 25 and 26. Fig. 25 illustrating the arrangement for securing the circulation, and Fig. 26 the method of building up the

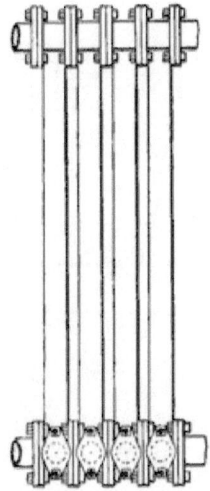

FIG. 26. METHOD OF CONSTRUCTING ECONOMISER OF SEPARATE TUBES.

Economiser in separate tubes. In Fig. 25 the arrows show the direction of the flow of the hot and cold water. In actual practice it was seldom found that either of the modifications of design here shown were necessary, but they are given as illustrating the attention that has been devoted to small details with a view to render the apparatus as perfect as care and the teachings of experience can possibly make it.

ECONOMISER BOTTOM BOXES

Sir E. Green Patent No. 2,989, 1887

In 1887, Patent No. 2,989, Sir E. Green proposed certain modifications in the construction of the bottom boxes with a view to the more efficient removal of any scale or sediment that might collect therein. The proposal consisted of the use of a mechanical rotating tool for loosening the sediment, combined with certain

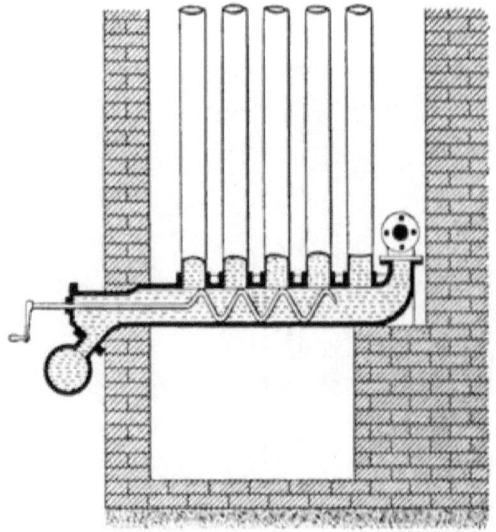

FIG. 27. ECONOMISER FITTED WITH AGITATOR FOR REMOVAL OF MUD AND SCALE.

modifications in the shape of the bottom box and mud chamber for the more easy removal of any accumulation by means of the blow-out. Various methods were proposed to effect this object. Fig. 27 shows one arrangement in which a mechanical worm or agitator was fitted in the bottom box, and revolved by hand, with a view to remove the accumulated mud or scale when the blow-out tap was opened.

Economiser with Feed Downflow

Sir E. Green Patent No. 8,651, 1895

The accompanying arrangements, Fig. 28, have been designed by Sir E. Green to prevent the vapour in the waste gases condensing on the outside of the Economiser tubes. This condensation or "sweating"

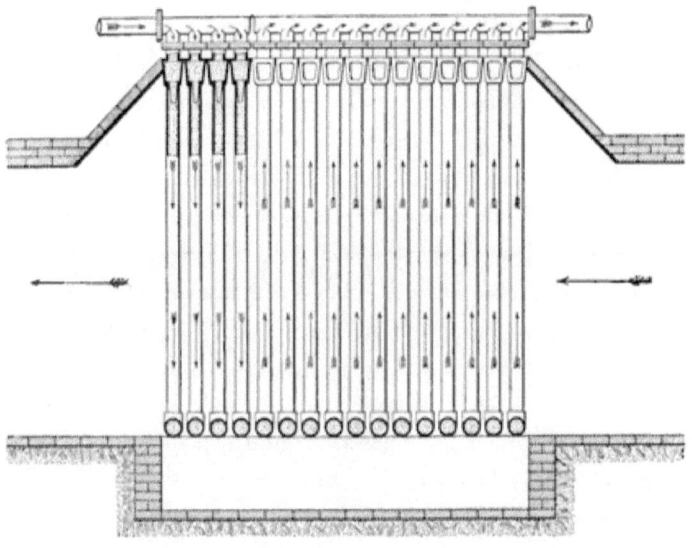

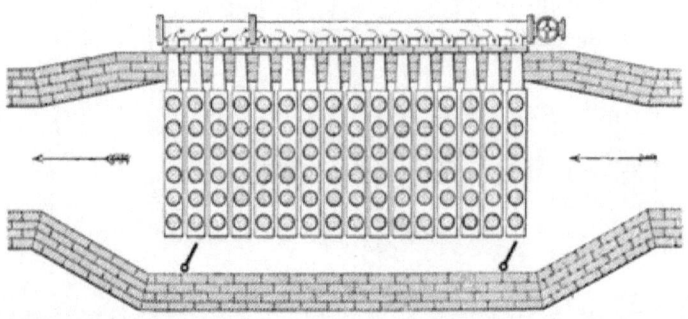

FIG. 28. ECONOMISER WITH FEED DOWNFLOW ARRANGEMENT TO PREVENT "SWEATING" OF THE TUBES

is most noticeable at the feed inlet end, especially at the lower part where the tubes are connected to the bottom boxes, and is objectionable as it combines with the sulphurous fumes given off by the coal, and forms an acid which rapidly eats away the tubes.

In the arrangement shown, the feed water instead of being delivered to the whole of the bottom boxes of the Economiser and flowing upwards in an uniform direction throughout the whole of the tubes, is delivered in the first instance to the top boxes of a number of sections which are isolated from the rest. This portion of the Economiser may consist of 4, 6, 8, or any number of sections as is found desirable. The upper end of each of the tubes in these sections is fitted with a small nozzle so as to restrict the area, and insure an approximately even distribution and downflow of the feed. By this means the inflowing current of feed water is raised in temperature before it reaches the bottom boxes, from whence it flows upwards through the remaining tubes in the usual way.

In order to further prevent the "sweating," the hot water outlet pipe of the Economiser and the suction side of the feed pump should be connected with a small pipe (say ⅜in. in diameter), to take off the chill whenever the temperature of the feed is 90° F. or less.

A CORNER OF THE BO SIDE.

OTHER ECONOMISER INVENTIONS

THE great economy that followed the application of Messrs. Greens' Economisers or Feed-water Heaters to steam boilers, led not only to its rapid and extensive adoption by steam users, but, as might naturally be expected brought a number of rival inventors into the field. The careful manner, however, in which every detail of the apparatus had been considered, left little room for improvement. Some of the rival designs proposed were utterly impracticable, and were never submitted to the test of actual work. Others appeared to possess good features, but when tested gave rise to so much trouble and difficulty, or developed such dangerous defects, that they were discarded after a short period of working. Space forbids enumeration of all the varied arrangements that have at different times been proposed, and their description would scarcely serve any useful purpose, though a brief reference to some of the more important ones may perhaps prove of interest.

The problem of keeping the surface of the Economiser tubes free from soot or other non-conducting coating was, as already pointed out, one to which Messrs. Green's attention was drawn, and to which considerable efforts were devoted at a very early stage in the Economisers' history, and subsequent experience

OTHER ECONOMISER INVENTIONS

under a variety of conditions has proved that the question of soot removal can only be effectively dealt with by means of mechanical scrapers clasping the outside of the tubes.

Lees Patent No. 1,325, 1864
In 1864, however, a Mr. J. W. Lees took out a patent, No. 1,325, in which he proposed to remove the soot by means of a series of fine steam jets, so arranged as to act as a kind of brush. The

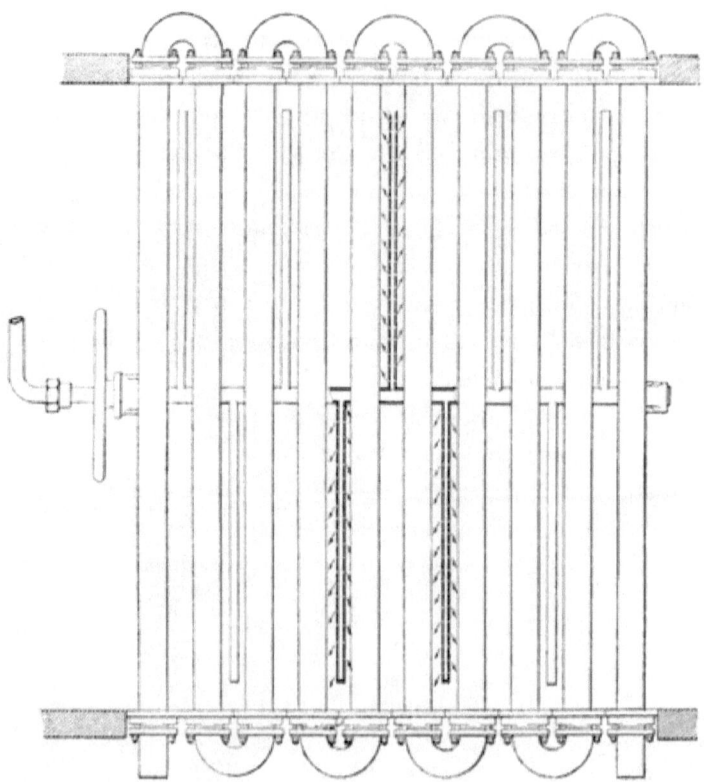

FIG. 29. LEES' REVOLVING STEAM BRUSH.

apparatus consisted of a steam pipe, which passed through the centre of the Economiser stack, and revolved in glands at each end. At intervals along the pipe hollow radial arms were fixed, having small perforations throughout their length. By turning a hand-wheel, fixed at one end of the central pipe, the radial arms described circles in the spaces between each sheet of tubes, and thus when steam was turned on the small jets played over the area of the circle described by the revolving arms. The general arrangement of this apparatus will be perceived on a reference to Fig. 29.

The method of cleansing the tubes, proposed by Mr. Lees, though apparently simple and promising well on paper, did not prove efficient in practice. It was found that the coating of soot became gradually thicker and of a hard adherent character, which slowly but surely prevented the passage of heat, and after a comparatively short time rendered the Economiser almost worthless.

In connection with this apparatus it may be noted in passing, that the tubes were arranged horizontally instead of vertically, while they were so connected as to cause the water to flow in one continuous stream from the inlet to the outlet, instead of slowly rising through the whole series of tubes, as in Messrs. Green's modern arrangement. This idea, however, of continuous circulation was not novel even thirty years ago, and the arrangement of the Economiser so as to insure the water traversing either each tube, or section of tubes in turn, was provided for in one of Green's earlier specifications.

Although the "continuous circulation" system may appear very attractive on paper, actual working has proved it on several occasions to be open to serious objection and even danger, while careful tests have shown that the continuous circulation Economiser does not possess any advantage on the score of economy. The matter, however, will be found more fully discussed on page 71.

Whitehead Patent No. 503, 1866

As an illustration of the manner in which the same ideas are re-invented, it may be stated that, in the year 1866, a Mr. J. H. Whitehead, in Patent No. 503, again proposed the use of a steam or compressed air brush. This was identical in principle with the one patented by Mr. Lees and described above, the only difference being that in the present patent the brush consisted of a series of pendant pipes which were traversed along the spaces between the Economiser tubes by means of a rack and pinion with the aid of a sliding stuffing box arrangement.

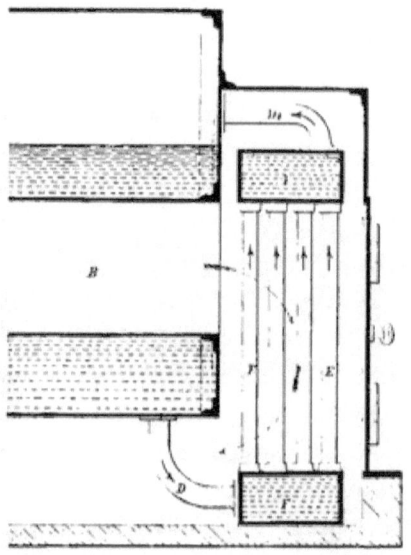

FIG. 30. SECTION OF TWIBILL'S SUPPLEMENTARY HEATER.

Twibill Patent No. 2,378, 1866

In the year 1866, Mr. Joseph Twibill took out a patent, No. 2,378, for improvements in the construction of supplementary Steam Generators and Scraping Apparatus. The principal feature in connection with Mr. Twibill's invention consisted in the position of the supplementary heater relative to the boiler and its connection thereto.

The form of the apparatus proposed varied somewhat with the type of boiler employed, but the general principle will be realised on a reference to Figs. 30 and 31, which show its application to a boiler of the ordinary Lancashire type. The heater consisted of a stack of tubes, E E, fitted with water-boxes, F F, at the top and bottom, and placed in the down-take at the back end opposite the furnace tubes, B B. The water-boxes, F F, were connected to the boiler by the tubes, D and D D, so that there was a continuous circulation through the supplementary heater, the water flowing from the boiler into the lower box, F, and thence passing up through the vertical tubes, E E, into the upper box any steam generated passing through the tube, D, into the boiler. This so-called "Feed-water Heater" was simply a water-circulating arrangement. As far as the temperature of the steam and water within it were concerned the apparatus was merely an extension of the boiler heating surface.

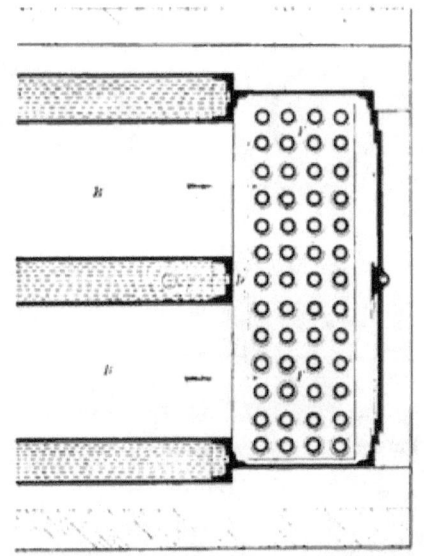

FIG. 31. PLAN OF TWIBILL'S SUPPLEMENTARY HEATER.

40 OTHER ECONOMISER INVENTIONS

Mannock Patent No. 2,086, 1867
In 1867, Patent No. 2,086, Mr. James Mannock proposed the arrangement of tube scraper shown in Fig. 32. The scraper, it will be seen, consisted simply of an annular cup or holder which encircled the tube, and carried within it two or three loose segments. These, by sliding

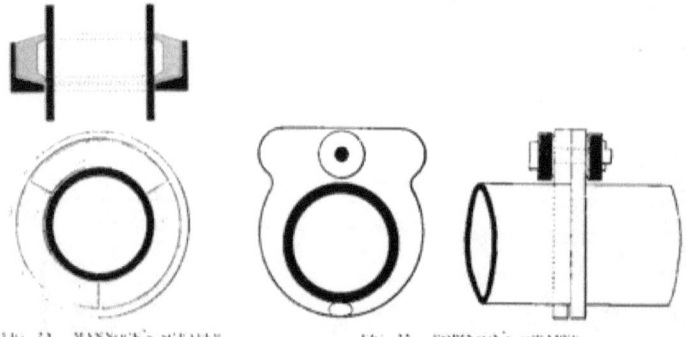

FIG. 32. MANNOCK'S SCRAPER.   FIG. 33. ROBINSON'S SCRAPER.

on the inclined bottom of the cup, were forced to bring their cutting edges against the outside of the tube and bear against it when making the up stroke.

Robinson Patent No. 679, 1868
Fig. 33 shows a proposal by Mr. J. Robinson, embodied in Patent No. 679, 1868, for scraping the outside of horizontal Economiser tubes. The scraper consisted of two parts which swivelled on a pin and clasped the outside of a tube like the legs of a pair of callipers. This arrangement, it will be obvious, would not permit of application to an Economiser of ordinary vertical tube construction.

Whitehead Patent No. 1,197, 1868
Another device for cleaning the soot off the outside of Economiser tubes was proposed, by the Mr. Whitehead already referred to, in Patent No. 1,197, 1868. The contrivance, however, was very crude,

and consisted simply of a series of short curved blades attached to chains which formed an endless band, and ran over pulleys at the top and bottom. The operation of these chains caused the blades to catch against the tubes—which were fixed horizontally—for a short portion of their circumference as it passed up and down on each side. A large portion of the surface of the tubes could, however, not be operated on at all by this device, and it is doubtful whether the arrangement proposed was ever tried in practice.

Holt & Galloway Patent No. 2,253, 1868 In Patent No. 2,253, 1868, Messrs. C. H. Holt and C. J. Galloway show a continuous circulation Economiser of the form illustrated in Fig. 34. In this arrangement the principle of "continuous

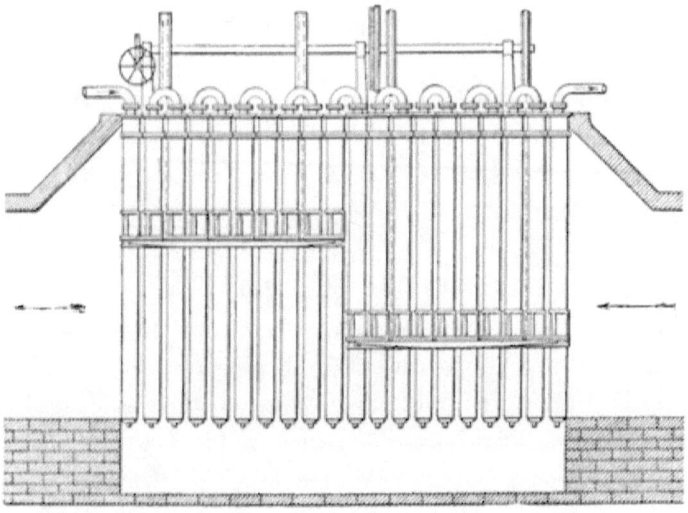

FIG. 34. HOLT & GALLOWAY'S CONTINUOUS CIRCULATION ECONOMISER.

circulation" was adopted, *i.e.*, the water was made to flow in one sinuous stream from the inlet to the outlet, the apparatus forming practically one continuous thoroughfare. The tubes were arranged vertically and hung pendulous from the top boxes. To ensure continuous circulation each tube was divided by a vertical diaphragm extending nearly, but not quite, to the bottom, the water flowing down one side and up the other, and so on in succession throughout the entire series of tubes from the inlet to the outlet.

A number of Economisers of this kind were put into operation, but the impossibility of locating the position of any block in the circulation arising from the deposit of sediment or other cause was found to give rise to serious trouble. Added to this, it was impossible to empty the Economiser except by boiling off the water, and in one case where this was attempted, the pressure, even with the safety valve open, increased to such an extent in consequence of the peculiar construction of the arrangement that a disastrous explosion resulted. This called wide attention to the dangerous character of this type of apparatus, unless equipped with a number of safety valves, and led to its being generally discarded.

Needham Patent No. 3,358, 1868 In the same year, 1868, a Patent No. 3,358 was granted to Mr. Richard Needham, for improvements in Economiser scrapers. The novelty claimed in this case was the fixing of the movable blades of the scraper, which were of flat semi-circular shape, in guides, so that the scrapers were inclined to the axis of the tubes at an angle of about 45°. The previous use of blades of the form described is admitted in the Patent Specification, and the only claim for novelty appears to have been the fixing of them at an inclination in the manner stated.

OTHER ECONOMISER INVENTIONS

Needham Patent No. 3,749, 1868

A few months later, Mr. Richard Needham, in Patent No. 3,749, 1868, described further designs for improving the operation of scraper driving and reversing apparatus. One method by which he proposed to operate the gear for working the scrapers is shown in Fig. 35. This consisted in placing within the Economiser tube an archimedean screw, having a central shaft revolving in a footstep at the bottom, and passing through a stuffing box at the top. The idea was, that the current of water travelling up the tube would act upon the screw and cause it to revolve, and that from the motion thus imparted to the shaft, the scrapers could be operated. The current of water flowing through an Economiser is so slow, however, that it is questionable whether any appreciable motion could be obtained in this way, and it is doubtful if the design was ever more than sketched on paper.

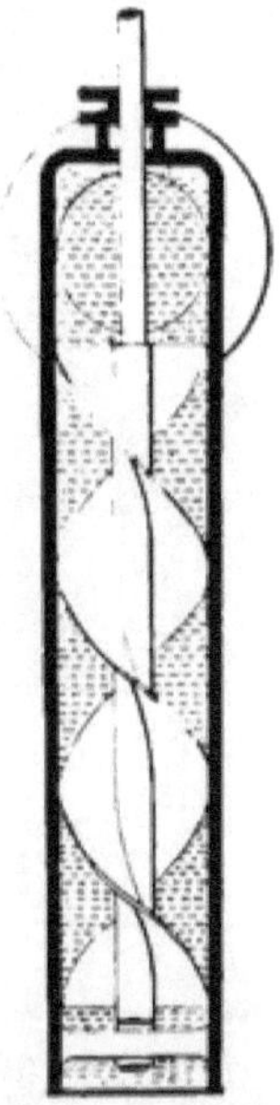

A more ingenious and practical proposal, taken from the same patent, is shown in Fig. 36,

FIG. 35. NEEDHAM'S SCREW SHAFT.

and represents a method of obtaining a reversing motion of the scrapers from the continuous revolution of a driving shaft. This shaft had cut upon it two screw threads—one right-handed, the other left-handed. Into these threads a shuttle, carried by a short sleeve, engaged, so that when the sleeve reached each end of the shaft the shuttle was transferred from the groove in which it was running to that of the opposite hand, and the sleeve thus carried alternately from end to end of the shaft.

To this sliding sleeve, chains led over pulleys were connected, the ends being attached to two sets of scrapers which balanced

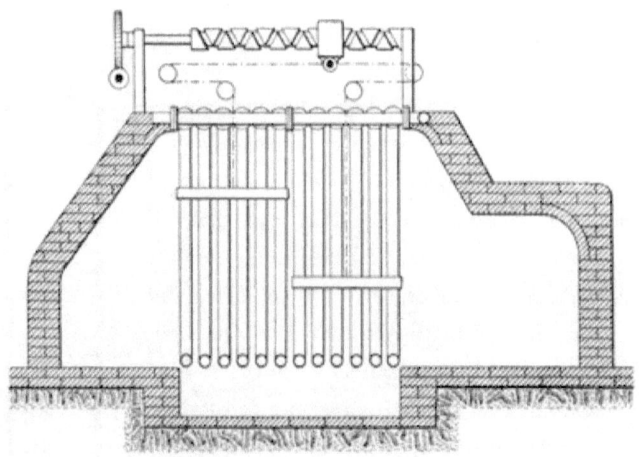

FIG. 30. NEEDHAM'S REVERSING MOTION FOR SCRAPERS.

each other, and moved up and down as the sleeve travelled to and fro.

In this same specification Mr. Needham suggested another improvement, which, however, like many others that have been proposed from time to time, is much easier to draw on paper than to carry out in practice. The proposal was that each sheet of vertical Economiser tubes, along with the top and bottom boxes, to which they were connected, should be all cast in one piece. However laudable this method of construction may be for some reasons, it is to be feared the difficulties of adopting it would prove insuperable.

OTHER ECONOMISER INVENTIONS

Obach
Patent No. 728,
1869

Fig. 37 shows a design which was proposed by Mr. T. Obach, in Patent No. 728, 1869, for securing Economiser tubes to the top and bottom boxes in such a way as to permit of their easy removal when necessary. The lower end of each tube was fitted with lugs on each side. These passed through a corresponding recess, which admitted of their access to a helical groove, so that by partially revolving the tube the spigot end was forced hard against a flat ring of copper or other soft metal placed on the joint face of the bottom box. At the upper end the tube was turned conical, and a corresponding hole bored in the top box, while the screw which held the cover in place was prolonged in the shank so as to bear against a cross-bar fitted in the upper end of the tube. The screw thus served the double purpose of bolting down the cover and forcing the tube into position. The design proposed was ingenious, but it would be much easier to put tubes together in this way when new than it would be to separate them after working for a time when the lugs would be rusted fast in the groove and almost impossible to release without fracturing the bottom box. Fig. 38 shows the arrangement of scrapers which was also covered by this specification. These were sector-shaped, and worked on a hinge, so that the scrapers permitted of a certain amount of play when they encountered any irregularity or obstruction on the face of the tube.

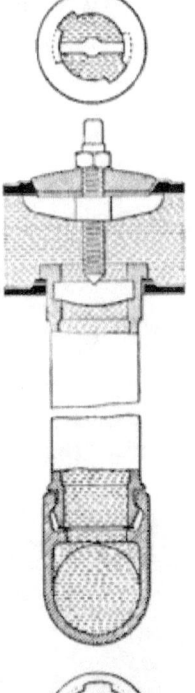

FIG. 37. OBACH'S TUBE JOINTS.

OTHER ECONOMISER INVENTIONS

The same patent included a design of the inventor for operating the reversing gear of scrapers. This is shown in Fig. 39. The driving pulley, A, actuates a worm, B, and thereby gives a constant revolution to the crank, C, attached to the worm wheel. By means of a connecting rod, D, a second crank, E, of larger throw is operated. This is attached to a toothed sector, F, which is thus caused to travel to and fro in the arc of a circle. Through the medium of a toothed wheel, K, which is fixed to the shaft carrying the scraper chain wheels, M, the latter are caused to alternately revolve in opposite directions, and thus raise or lower the scraper frames to which they are attached.

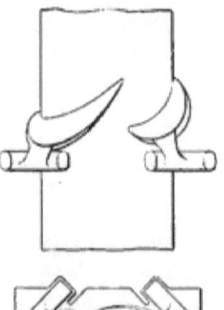

FIG. 38. ORACH'S SCRAPER.

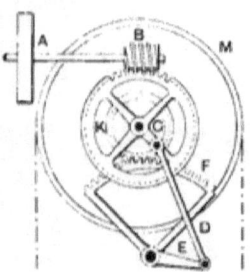
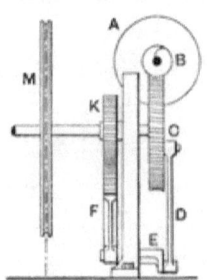

FIG. 39. ORACH'S REVERSING MOTION FOR SCRAPERS.

OTHER ECONOMISER INVENTIONS

Bell Patent No. 3,365, 1870

In 1870, Patent No. 3,365, Mr. Andrew Bell proposed to arrange the Economiser tubes in the form of a series of spirals similar to the one shown in the accompanying sketch, Fig. 40, instead of in vertical rows as in Messrs. Green's arrangement. The tubes were kept clean on the outside by the action of three radial scrapers which were operated by the central vertical shaft, B, and travelled down the helical tube until they reached nearly to the bottom, when, by means of the reversing gear the direction of rotation of the shaft B was reversed. It is difficult to see what advantage could be reaped by arranging the tubes spirally, while such an arrangement absolutely prohibited the cleaning of the tubes internally. This, apart from other difficulties, constituted a fatal objection, and although the apparatus was practically tried, it had an exceedingly

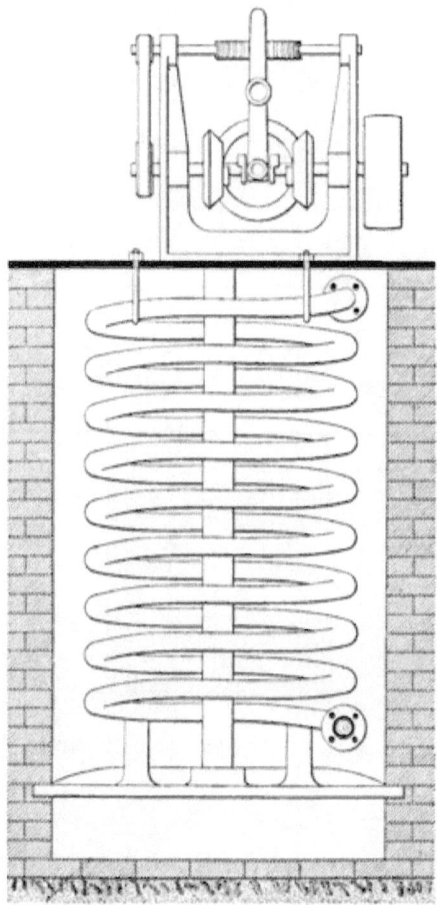

FIG. 40. BELL'S SPIRAL ECONOMISER.

E

limited application, and the trouble experienced in its working soon caused it to be abandoned.

Mundy Patent No. 2,876, 1871
Mr. E. Mundy, in Patent No. 2,876, 1871, proposed certain improvements in the design of scrapers. In this case the scrapers consisted of two sets of segments encircling the tube and placed one above the other, the segments being operated by a ring or hoop having a bevelled edge, so that a pressure could be exerted when the scraper frame was raised in the usual way.

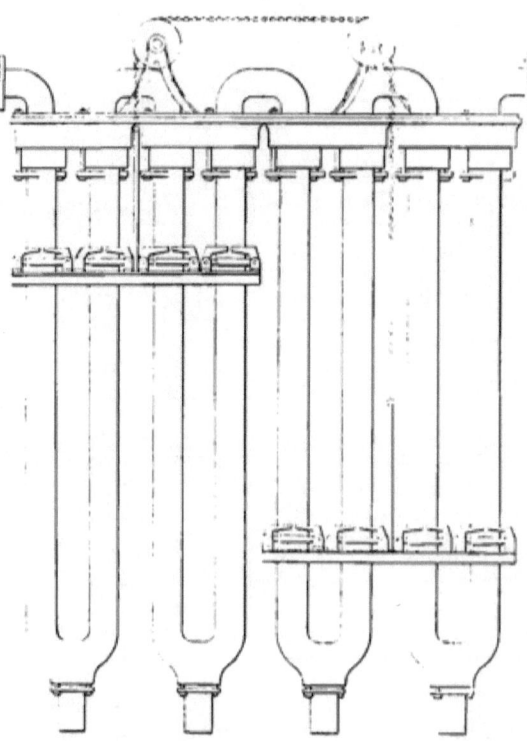

FIG. 41. ELEVATION OF CALVERT & TAYLOR'S CONTINUOUS CIRCULATION ECONOMISER.

OTHER ECONOMISER INVENTIONS

Calvert & Taylor Patent No. 350, 1872

In 1872, the principle of continuous circulation was again "invented" and made the subject matter of Patent No. 350 granted to Messrs. J. C. Calvert and J. Taylor, of Huddersfield. The arrangement proposed by them is shown in Figs. 41 and 42. The tubes it will be seen were vertical, and depended from a series of top boxes in pairs, each pair being connected at the bottom with a semi-circular bend, while the top boxes were divided with diagonal partitions, so

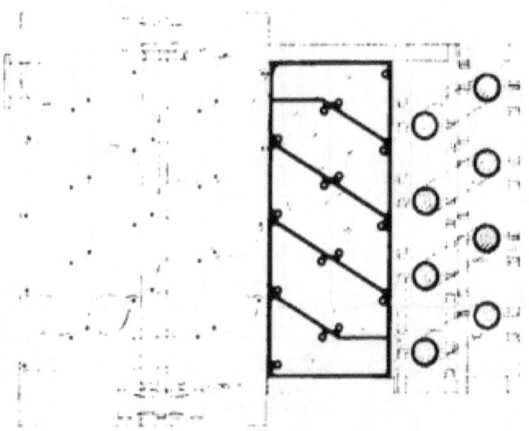

FIG. 42. PLAN OF CALVERT & TAYLOR'S CONTINUOUS CIRCULATION ECONOMISER.

that the tubes belonging to each pair were connected to a separate compartment. These were so arranged that the water was caused to travel in a continuous stream through the whole series of tubes, from the inlet to the outlet.

Elson Patent No. 695, 1872

The accompanying illustrations, Figs. 43 and 44, show an arrangement of Economiser devised by Mr. S. Elson, Patent No. 695, 1872. The apparatus consisted of two rows of tubes, or "bottles" as they were termed, the lower ends being blanked up and the upper ends of each row connected to one longitudinal collecting pipe. The water entered the Economiser through a small tube placed inside the upper longitudinal one, and from this flowed down a central tube, which extended nearly to the bottom of each "bottle," the ascending

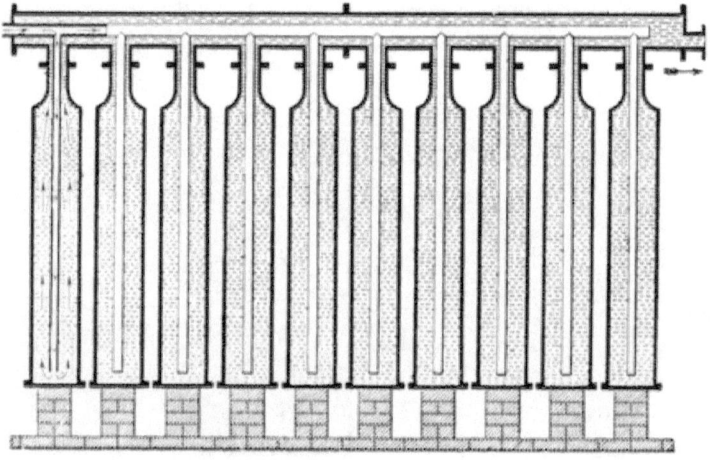

FIG. 43. ELSON'S BOTTLE ECONOMISER.

currents of water being again collected in the upper pipe, and thence passed to the boiler. The scrapers were of a somewhat novel design, consisting of revolving strips of wrought-iron, which were bent helically round the "bottle," and connected to a loose collar at the bottom and a worm wheel at the top, which was operated by a central shaft passing between the two rows of "bottles." See Fig. 44.

A number of these Economisers were put to work, but the grave defects which attended their working, and which in one or two instances resulted in disastrous explosions, led to the complete abandonment of the type in the course of a few years. In the case of muddy or dirty water there was no means of blowing off the Economiser, and the accumulation of dirt and sediment at the bottom choked the end of the flow-pipe, so that the water in the "bottle" became stagnant or idle. In the event of the upper neck becoming choked as well, which it was very liable to do on account of the narrow water space, the "bottle" was at once converted into an hermetically-sealed vessel, and on the application of heat explosion became almost inevitable. The pendant form of the "bottles," coupled with their large diameter, was also a source of weakness, and in the case of one "bottle" bursting, the others followed in succession like a row of skittles. Further, from an economical point of view, the large diameter of the "bottles" was a defect, since the ratio of heating surface to capacity is less with large tubes than with small ones.

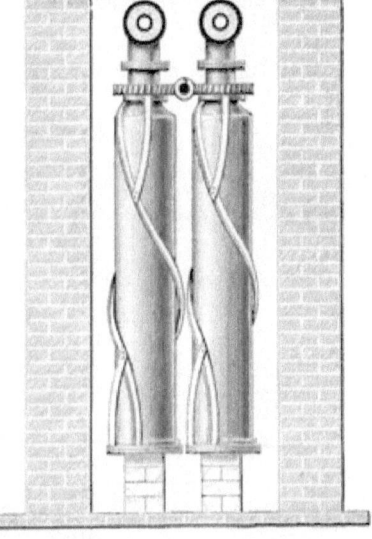

FIG. 44. ELSON'S SPIRAL SCRAPERS.

Lees & Garforth Patent No. 3,324, 1872   Messrs. R. J. Lees and J. H. Garforth, in Patent No. 3,324, 1872, proposed certain improvements in the form of scrapers as well as in the arrangement of the flow of water through the tubes. The design of scraper proposed by them is shown in Fig. 45. The scrapers consisted of flat-bars, running lengthwise of the tube, and attached to iron rods, which were connected to a revolving plate at the upper end. These plates were secured to a series of wheels which geared

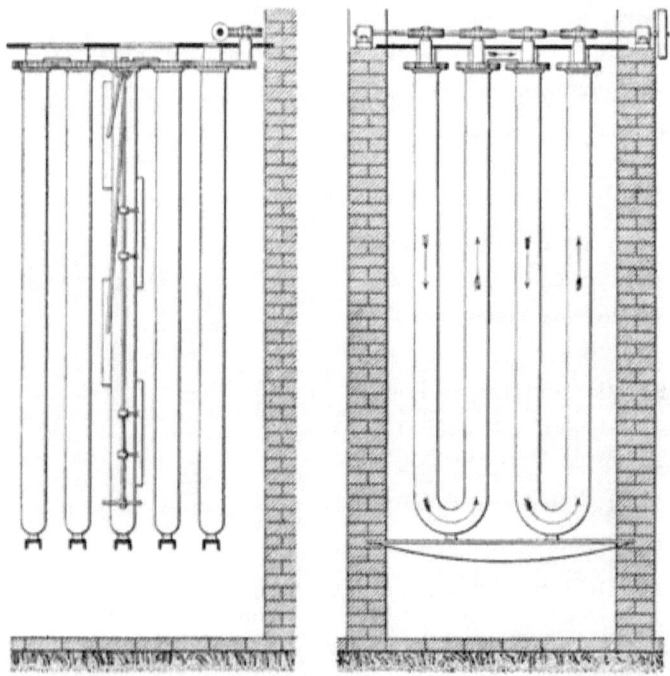

FIG. 45. ILES & GARFORTH'S CONTINUOUS CIRCULATION ECONOMISER

into each other so that the whole of the scrapers in a sheet of tubes were driven together and operated by a single worm and wheel. It will be observed that the tubes were scraped in a circumferential and not in a longitudinal direction, as in nearly all the other arrangements that had been proposed.

In this apparatus the tubes were cast in pairs, united with a U-shaped bend at the bottom, and connected with one another

OTHER ECONOMISER INVENTIONS

by elbows at the top, in such a way as to cause continuous circulation. In the provisional specification this was claimed as

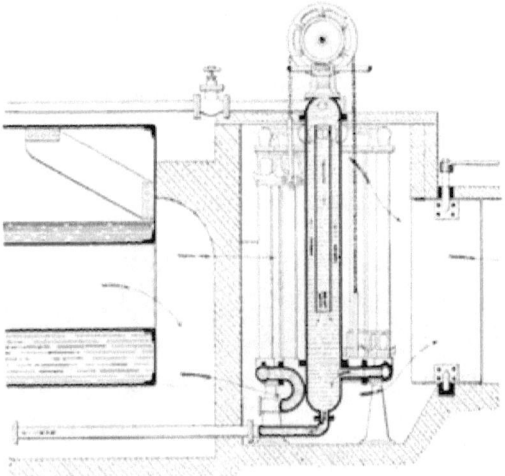

FIG. 46. ELEVATION OF TWIBILL'S CIRCULAR ECONOMISER.

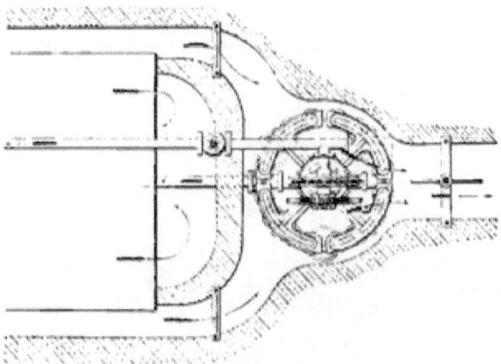

FIG. 47. PLAN OF TWIBILL'S CIRCULAR ECONOMISER.

part of the invention, but in the complete patent the idea of novelty on behalf of the circulation of the water was repudiated and no claim made in respect of it.

**Twibill
Patent No. 3,791,
1873**

In the year 1873, Mr. J. Twibill took out a Patent, No. 3,791, for an Economiser of the form shown in the illustrations, Figs. 46 and 47. The main feature of Mr. Twibill's invention consisted in arranging the Economiser tubes in one or more circles attached to a central water chamber, the advantages claimed on behalf of the arrangement being that greater facility was afforded for dealing with filtrated or precipitated sediment.

**Mason & Alcock
Patent No. 1,436,
1874**

In 1874, the principle of continuous circulation was once more brought forth in Patent No. 1,436, granted to Messrs. S. Mason and M. Alcock. The arrangement proposed is shown in Fig. 48. The tubes in this

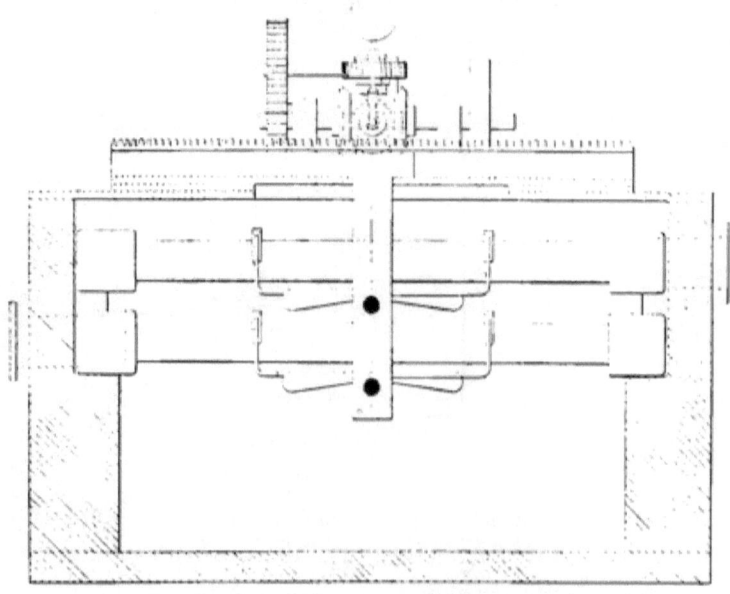

FIG. 48. MASON & ALCOCK'S CONTINUOUS CIRCULATION ECONOMISER.

design were arranged horizontally, and fixed in two tiers, one above the other, which ran transversely across the flue, the scrapers

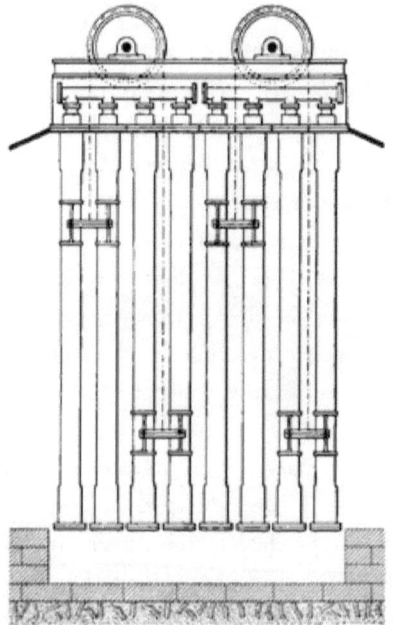

FIG. 49. ELEVATION OF PRESTWICH & PIMBLEY'S SPLIT TUBE ECONOMISER.

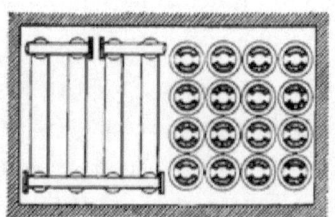

FIG. 50. PLAN OF PRESTWICH & PIMBLEY'S SPLIT TUBE ECONOMISER.

being operated by means of a sliding rack and pinion, in conjunction with an ordinary reversing gear.

Prestwich and Pimbley Patent No. 4,235. 1874

Figs. 49 and 50 represent an idea embodied in Patent No. 4,235, 1874, by J. Prestwich and J. Pimbley, with the object of increasing the surface of Economiser tubes and diminishing the volume of water. It was proposed to effect this by constructing the body of the vertical tube in two segments, as shown in the plan, Fig. 50, the space between the segments permitting of the introduction of a scraper so that soot could be removed both from the convex and concave sides. In this, as well as in other similar devices that have been proposed for diminishing the volume of water in the Economiser, inventors appear to forget that by reducing the volume of water in the apparatus its rate of flow is necessarily accelerated, so that, although the heating surface is relatively greater, the length of time the water remains in contact with it is correspondingly curtailed, and what is gained in one direction is lost in another. In the construction, as sketched in this Patent, it will be observed that there is no arrangement for insuring the circulation of water through the tubes, which consisted of a series of bottles of dead water with blind ends at the bottom. The feed inlet, as well as the delivery outlet, was situated at the top, and the efficiency of an apparatus so constructed would prove very small.

Bell Patent No. 639, 1880

In Patent No. 639, 1880, granted to Mr. Andrew Bell, several so-called novelties are described and are here illustrated in Fig. 51. The first claim made is for the coupling together of the Economiser tubes by a series of elbows at the top and bottom, so that each transverse row is in the form of a gridiron, while each row is connected to the succeeding one in such a way that the water is made to flow in one continuous stream throughout the whole length of the Economiser

A CORNER OF THE FITTING SHOP

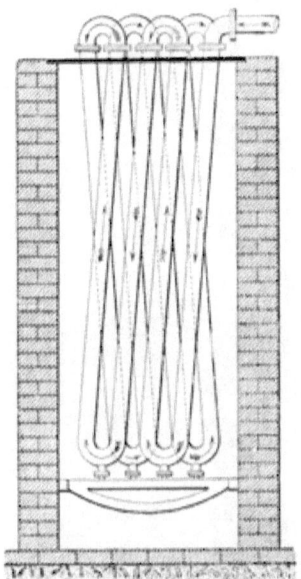

FIG. 51. BELL'S CONTINUOUS CIRCULATION ECONOMISER.

from the inlet to the outlet. This is the Continuous Circulation Economiser again. A second claim made by Mr. Bell is that the water passes through the Economiser in a direction contrary to the flow of the heated gases. This was adopted in Green's earliest patents. A third claim was for the inclination of the separate rows of tubes in opposite directions, with the object of causing the gases flowing through the spaces in one row to impinge on the tubes in the next, but as the preceding pages show, this idea had been repeatedly exploited.

Sykes Patent No. 1,746, 1882 The next invention proposed for the improvement of the Economiser was one by Mr. T. Sykes, in 1882, Patent No. 1,746. The invention consisted in placing three or four tubes together so as to form a group, as shown in Fig. 52, these groups being connected to longitudinal or transverse boxes at top and bottom by a tapered spigot in the usual way. The object of this invention, as stated in the specification, was " to obtain an increased amount of heating surface and a more rapid circulation of the water." It is difficult to see how this end could be attained by the means proposed. The grouping of the tubes together would certainly tend to diminish the heating surface rather than increase it, and the circulation could not be in any way effected as

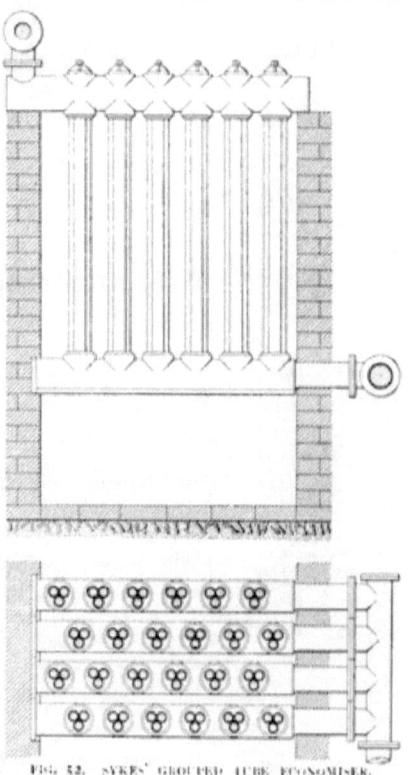

FIG. 52. SYKES' GROUPED TUBE ECONOMISER.

compared with an Economiser of ordinary construction, provided the volume of water remained the same.

Lowcock & Taylor Patent No. 3,306, 1882   In Patent No. 3,306, 1882, Messrs. A. Lowcock and J. Taylor suggested certain improvements in the mechanism for raising and lowering the scraper frames of Economisers, as well as the use of sheet-iron blades beneath the scrapers for passing between the transverse bottom boxes and preventing any accumulation of soot choking the thoroughfare. As this patent refers rather to matters of a detailed character, it scarcely calls for further description.

OTHER ECONOMISER INVENTIONS 59

Perkins & Scott Patent No. 3,502, 1882

In 1882, Messrs. J. G. Perkins and J. Scott took out a Patent, No. 3,502, referring principally to the construction of scrapers and of internal tapered access lids for the top boxes of Economisers, thereby dispensing with the use of bolts and cross-bars. The scrapers were of what is known as crab-jaw construction, and consisted of circular segments, embracing a portion of the tube, and connected by levers to a lug or hinge on the scraper frame, the action of which caused the scrapers to exert pressure at the upper or lower ends according as they were raised or lowered.

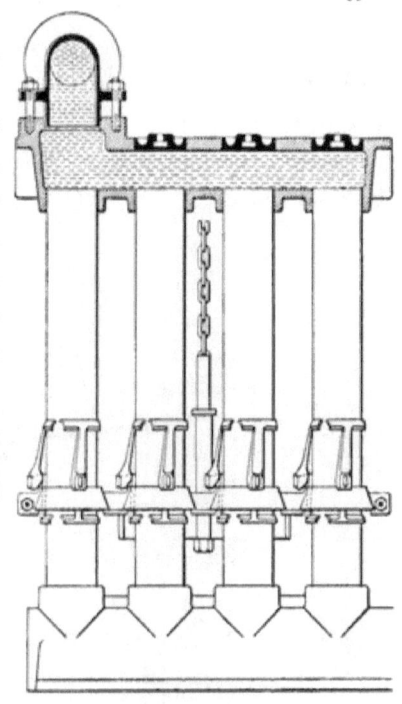

FIG. 53. PERKINS & SCOTT'S CRAB-JAW SCRAPERS.

Lowcock & Sykes Patent No. 2,875, 1885

The illustrations given in Figs. 54 and 55 show certain arrangements for blowing-out Economisers, which were proposed in Patent No. 2,875, 1885, by Mr. A. Lowcock and Mr. T. Sykes. The claim for novelty consisted in constructing the bottom box in such a way as to cause it to slope towards the blow-out pipe, either as shown in Fig. 54 where the blow-out pipe runs longitudinally through the soot chamber, or as shown in Fig. 55 where the bottom box is divided into two compartments by a

60   OTHER ECONOMISER INVENTIONS

sloping diaphragm extending nearly throughout its length, the blow-off pipe being connected to the lower compartment. By this means it was contended that any mud or dirt accumulating in the bottom box could be more efficiently cleared away on opening the blow-out tap.

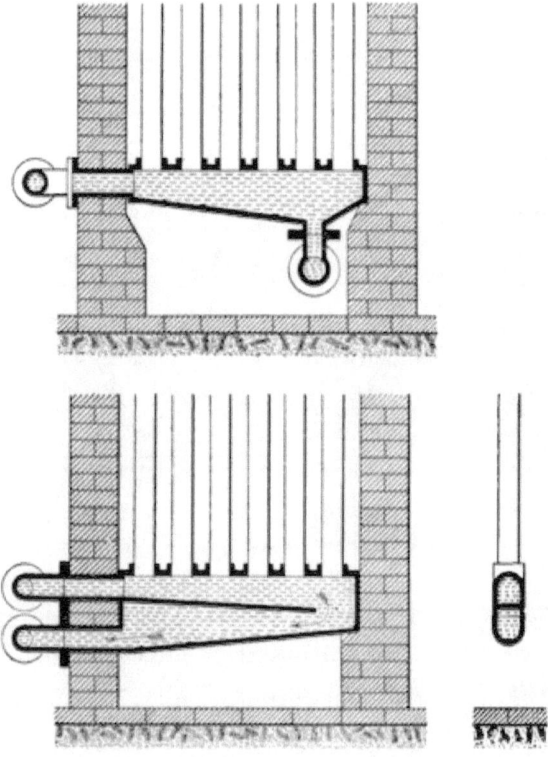

FIGS. 54 & 55.   LOWCOCK & SYKES' BLOW-OUT ARRANGEMENTS.

Bell Patent No. 8,020, 1885   In 1885, Mr. A. Bell, in Patent No. 8,020, described an arrangement of Economiser shown in the accompanying sketches, Figs. 56 and 57, with a view to obtaining a continuous circulation of water, while

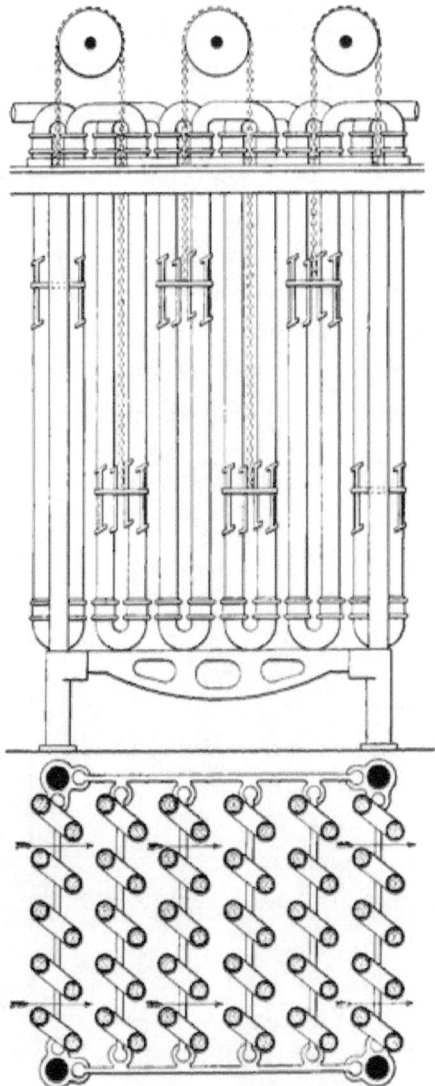

FIGS. 56 & 57. PLAN AND ELEVATION OF BELL'S CONTINUOUS CIRCULATION ECONOMISER.

the tubes were arranged in staggered rows, with the object, it was claimed, of causing the heated gases to impinge more directly upon their surfaces than was possible when they were arranged in line behind each other.

It is scarcely necessary to discuss the details which Mr. Bell proposed to insure continuous circulation, as a reference to the abstracts already given will show that the idea was in no sense new, and as regards the arrangement of the tubes in staggered rows for securing a more direct impingement of heated gases upon them, it may be pointed out that this arrangement was shown in Green's first patent, see Fig. 4, page 12. The reasoning, however, as to the superiority of this plan is to a great extent fallacious. The absorption of heat by Economiser tubes is determined not by the impingement of the gases upon them, but by the difference in temperature between the gases outside and the water inside. In other words, the amount of heat taken up by each tube is influenced by this temperature difference rather than by the position of the tube with respect to the gases flowing past it.

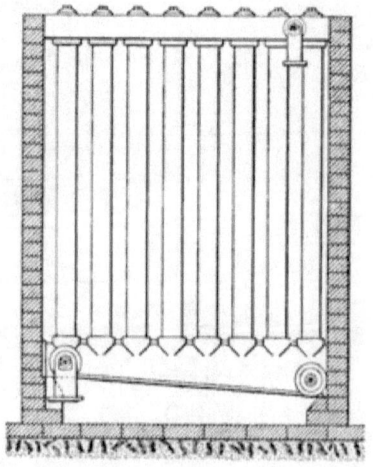

FIG. 58. HAWKINS' TOP AND BOTTOM BOXES.

Hawkins Patent No. 14,141, 1887 In Patent No. 14,141, 1887, Mr. G. C. Hawkins proposed to do away with the longitudinal branch tubes, used for connecting the top and bottom boxes together, in the manner shown in Figs. 58 and 59. In this arrangement the top and bottom boxes were to be provided with facings around a series of holes, which permitted of them communicating with each other, the end boxes of the series being

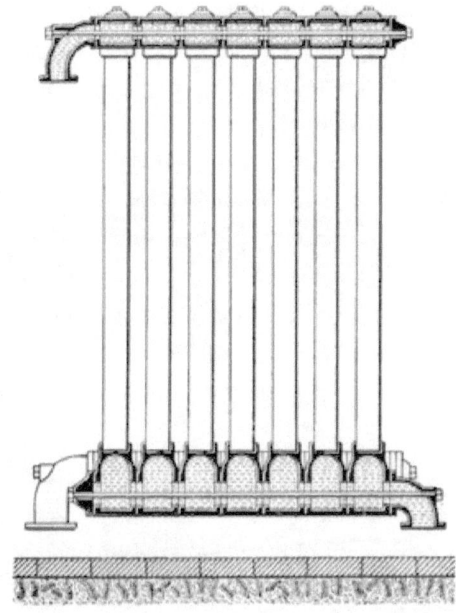

FIG. 59. HAWKINS' TOP AND BOTTOM BOXES.

fitted with an elbow and flange for connecting to the feed inlet, feed outlet, or blow-out pipe. For the purpose of drawing the various sections together and of resisting the tendency to be forced apart by the internal pressure, bolt stays were provided as shown in Fig. 59. Any advantages that can possibly accrue from such an arrangement as this are more than counterbalanced by the practical difficulty which would be experienced in making the joints.

F

Lowcock & Sykes Patent No. 16,114, 1887 Judging from the number of patents that have at one time or another been taken out with a view to promoting the flow of water in Economisers, it would appear that those inventors who have taken up the subject entirely overlook the fact that in the Economiser, as constructed by Messrs. Green's, the circulation in each tube is regulated by the amount of heat it receives. If one tube receives more heat than another the result is a greater amount of convection and a greater inflow of cold water, the circulation throughout the Economiser thus adjusting itself in a natural manner. Where there is most heat there is the most rapid circulation, and where there is least heat there is the slowest circulation. The distribution of heat and circulation go hand in hand and mutually accommodate each other. From this it will be evident that designs to make the water flow at a constant rate through every tube in an

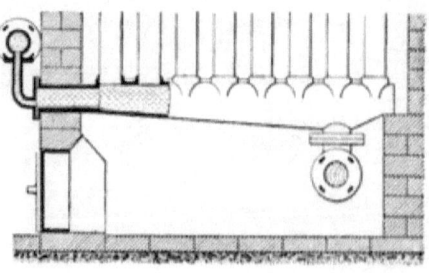

FIG. 60. LOWCOCK & SYKES' BOTTOM BOXES

Economiser with a view to promote economy, are futile. Hence such a proposal as the one shown in Fig. 60, which was the subject of Patent No. 16,114 granted to Messrs. A. Lowcock and T. Sykes in the year 1887, is of doubtful value. The patent consisted simply in making the bottom boxes of a taper-shape with a varying diameter proportional to the number of pipes to be fed.

OTHER ECONOMISER INVENTIONS

Twibill Patent No. 7,354, 1888

Mr. Joseph Twibill, in 1888, Patent No. 7,354, suggested an arrangement of Economiser with internal circulating tubes, as shown in Fig. 61. Each top box of the Economiser was divided into two compartments by a horizontal diaphragm, extending from end to end, and forming a tube plate for fixing the internal circulating tubes. The

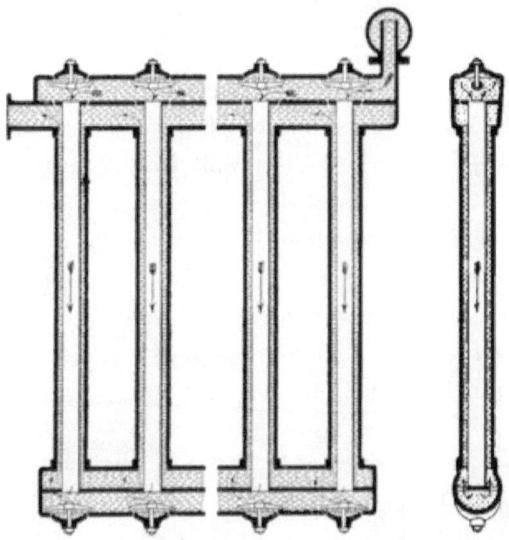

FIG. 61. TWIBILL'S INTERNAL CIRCULATING TUBE ECONOMISER.

bottom box was also fitted with a tube plate, but this was only of bar-shape, having a slight enlargement at the parts where the circulating tubes were fixed, and did not divide the bottom box into two separate compartments as in the case of the upper one. The circulation of the water is fairly indicated by the direction of the arrows in the accompanying sketches. From the feed inlet the water passed into the upper compartment of the top box and thence downward through the inner circulating tubes to the bottom box from which it flowed upwards through the series of annular spaces between the inner circulating tubes and the Economiser tubes proper into the lower compartment of the top box, and then away to the outlet. It is difficult to see what practical advantage could accrue from this arrangement, as the heating surface is in no way increased, and the internal cleaning of an Economiser of this kind would present very serious difficulties.

Sykes Patent No. 4,778, 1890

The accompanying illustration, Fig. 62, shows the construction of caps or lids of Economiser tubes which forms the basis of Patent No. 4,778, granted to Mr. T. Sykes in 1890. The cap is provided with a circular rib of angular section, which fits into a corresponding groove or seating turned in the face of the top box. This, it is claimed, permits of the joints being broken with less risk of injury to the top box. The joint is made either by means of lead putty in the usual way, or by using a ring or lining of copper or other suitable metal laid in the bottom of the circular groove referred to.

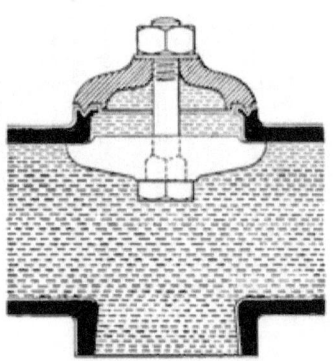

FIG. 62. SYKES' ECONOMISER LID.

Burpee Patent No. 21,243, 1890

In Patent No. 21,243, 1890, taken out by G. H. Burpee, several claims for novelty in the construction of Economisers are made. These relate principally to the use of spigot joints instead of flanges for connecting the bottom and top boxes to the branch pipes, and also to the method of forming the taper of the spigot ends of the vertical tubes. The novelty in respect to this latter claim consisted in making the taper of the ends of the Economiser tubes point downwards in each case, with the object of more easily withdrawing and renewing a tube in the event of repair. The idea is one, however, that presents greater advantages on paper than are found to exist in practice.

Sykes Patent No. 6,439, 1891
A few months after Mr. Burpee's patent, just described, was granted, another patent, No. 6,439, 1891, was taken out by Mr. T. Sykes having an identically similar object, namely, the formation of the tapers on the ends of the Economiser tubes so that the tapers pointed downwards in each case.

Topham Patent No. 13,438, 1892
In the year 1892 Mr. E. Topham took out a patent, No. 13,438, for connecting two vertical sheets of tubes to a common top and bottom box, the vertical tubes being so arranged that those in one row were opposite the spaces of the next row, with the object,

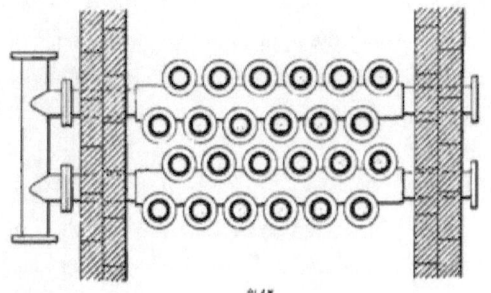
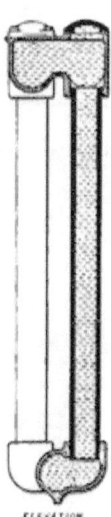

FIG. 63. TOPHAM'S DOUBLE TUBE BOXES.

it was claimed, of rendering the heating surface of the tubes more efficient by causing the stream of gases to impinge more directly upon them. Whatever advantage may be supposed to accrue from this, and it is very doubtful whether, as already pointed out on page 62, any possible economy does result, the proposal to arrange the tubes in staggered rows instead of in parallel lines was at all events not novel. The arrangement of the tubes in this way was shown in Mr. Edward Green's first patent, see Fig. 4, page 12.

Sankey Patent No. 21,918, 1892   Patent No. 21,918, 1892, refers to a design of feed-water heater proposed by Mr. W. H. Sankey. The arrangement described in the specification consists of a series of tubes connected to flat chambers at the top and bottom, so arranged as to cause a continuous flow through the whole series from the inlet to the outlet, and refers more particularly to heating by exhaust steam, though the claim is also made for its use in the main flue of a boiler to utilize the waste heat of the gases. Before it could be adopted, however, in such a position, the design as it is described would require very extensive modification, and is merely referred to here as an illustration of the extent to which the application of the principle of continuous circulation to the Economiser has been made the subject of letters patent.

Calvert Patent No. 4,775, 1893   From the references that have been already made, it will be apparent that the principle of continuous circulation in Economisers has often been the subject of investigation by inventors of Economisers, and it might be thought that the capacity for taking out patents in this direction would have become exhausted. Another patent for an Economiser of this type was granted to Mr. J. G. Calvert in 1893, Patent No. 4,775. As regards principle, this apparatus is identical with that of previous ones, the tubes being vertical and connected so as to cause the water to flow alternately up and down the various sheets of tubes in succession from the inlet to the outlet. A number of minor arrangements of the taps and valves are covered by the patent. These appear, however, to give rise to great complication.

Sykes Patent No. 5,401, 1894   This patent refers to a method of securing the lids of the holes in the top boxes. The lids are constructed in the form of screwed plugs, with a faced seating or bearing surface fitting into correspondingly screwed nipples formed on the upper side of the top boxes. It would no doubt be an easy and simple matter to secure the lids and make the joints in this way; but after they have been in position for some time and exposed to the action of heat and water, the threads of the lid and nipple, it is to be feared, would become rusted so fast together as to render it next to impossible to separate them. It may further be pointed out that the construction of screw lids was covered by Mr. E. Green's Patent, No. 3,794, in 1882. See page 25.

Sykes Patent No. 13,409, 1894   In 1894, Mr. T. Sykes, in Patent No. 13,409, proposed to interfere partially with the thoroughfare of the vertical tubes of Economisers by inserting in each tube one of smaller diameter, either blind or partially blind at its upper end, and projecting into the bottom box. By this means it is claimed the water in its passage through the

Economiser will be compelled to traverse the annular space between the inner and outer tubes, and thus become heated " more rapidly and to a higher temperature." The misapprehension that underlies several of the inventions of this class that have been proposed has been already so fully referred to that it scarcely calls for further reference here. Inventors appear to forget when drawing up such designs, that if the volume of water flowing through the Economiser is reduced, the average length of time occupied in its passage is correspondingly shortened, and thus any increase in the efficiency of the heating surface is neutralised by the diminished length of time the water remains in contact with it. The gain in one direction is balanced by the loss in another. We would further add that this patent appears similar to No. 877, taken out by Mr. E. Green and Mr. E. Green, Junior, in 1858. See page 16.

Knight & Thode Patent No. 24,320, 1894 This joint patent of C. A. Knight and G. W. Thode refers to a feed-water heater consisting of several groups of horizontal or inclined tubes, connected at the ends to vertical headers, these being divided in such a way that the water circulating through them is caused to flow in succession through the several sections of the apparatus. It is difficult in this arrangement to find any features of novelty. The use of horizontal or slightly-inclined tubes, as will be seen on reference to previous sketches, is in no sense new, any more than the principle of continuous circulation. In the arrangement shown in the specification the tubes are unprovided with scrapers; and there can be little doubt such an apparatus as the one sketched would, in practice, prove very inefficient in consequence of the deposit of soot.

SCRAPER GEAR FITTING ROOM.

ECONOMISERS
HAVING
CONTINUOUS CIRCULATION

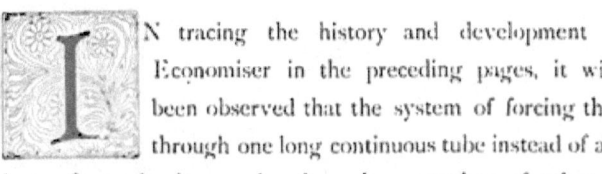N tracing the history and development of the Economiser in the preceding pages, it will have been observed that the system of forcing the water through one long continuous tube instead of allowing it to flow simultaneously through a number of tubes, as in Messrs. Green's system, has been the subject of many patents. It has, in fact, been designed over and over again.* Inventors appear to have been impressed with the idea that a continuous flow, by causing the water to pass through one long tortuous length of tube instead of a number of short ones, must result in its being heated to a greater extent, forgetting that such an arrangement necessitates the water travelling at a correspondingly higher velocity, and that for a given rate of feed the time occupied in passing through the Economiser is precisely the same whether the apparatus be arranged on the continuous system or in accordance with Messrs. Green's arrangement, and, further, that the amount of heating is determined by the length of time the water is in contact with the gases, not the velocity with which the water travels.

---

* *For inventions referring to continuous circulation see following—*

Lees' Patent, No. 1,325, 1864, page 36
Holt and Galloway's Patent, No. 2,253, 1868, page 41
Calvert and Taylor's Patent, No. 350, 1872, page 49
Lees and Garforth's Patent, No. 3,324, 1872, page 51
Mason and Alcock's Patent, No. 1,436, 1874, page 54

R. H's Patent, No. 639, 1880, page 56
Bell's Patent, No. 8,020, 1885, page 60
Sankey's Patent, No. 21,918, 1892, page 68
Calvert's Patent, No. 4,775, 1893, page 68
Knight & Thode's Patent, No. 24,320, 1894, page 70

This consideration alone is sufficient to show the fallacy of the views underlying the designs of Circulating Economisers, while it may be pointed out that in such arrangements the water encounters greater resistance in the Economiser in consequence of its higher velocity, and therefore calls for more pumping power. Further, unless special provision is made, there is in some designs of Circulating Economisers a risk of the pressure, under certain conditions, becoming excessive. In fact, one of the most disastrous Economiser explosions on record was due to the apparatus being arranged on the continuous circulating principle.

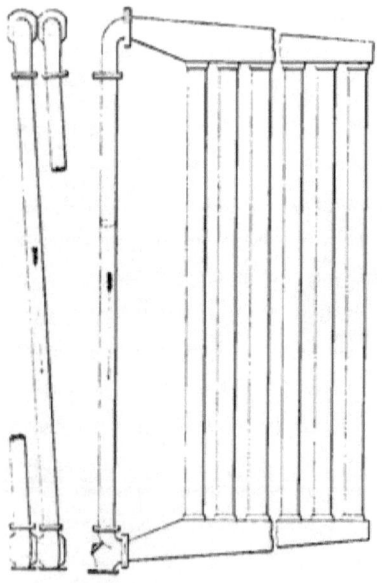

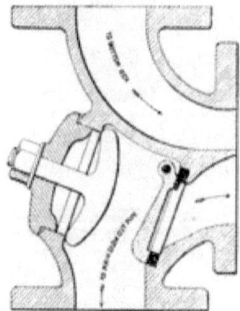

FIG. 64. CALVERT'S ECONOMISER; METHOD OF CONNECTING SECTIONS.

FIG. 65. BLOW-OUT ARRANGEMENT OF CALVERT'S ECONOMISER.

During the last year or two strenuous efforts have been made to revive the circulating type of Economiser. With a view to avoid the risk of excessive pressure which existed in early arrangements, this latest type of the apparatus has been equipped with safety valves on each section, while the blow-out arrangement has necessitated the use of a back-pressure valve, for each bottom box, the valve being either of ball type or the simple flap form shown in the accompanying illustrations, Figs. 64 and 65.

These various valves, it will be evident, render the apparatus complicated, and, from an economical point of view, do not give the slightest advantage. As extraordinary claims, however, have been made by the Patentees of this apparatus on the score of economy, it was thought desirable to submit the two types to a series of thoroughly impartial and independent tests by engineering observers whose character and professional standing should be beyond question, and with this object the Chief Engineers of the four leading Boiler Insurance and Inspecting Companies in the country were requested to conduct the tests.

The results obtained at these trials were unanimous and conclusive. They fully confirmed previous experience, and proved that, as far as economy is concerned, the Circulating Economiser is in no degree superior to the ordinary design, but, on the other hand is liable to cause trouble from its complicated arrangement.

In support of this, the following extracts from the Reports of the Chief Engineers who made the tests, are quoted:—

J. F. L. Crossland, M. Inst. C.E., the Chief Engineer of the Boiler Insurance and Steam Power Company, Limited, Manchester, for instance, at the conclusion of his report, dated September 7th 1894, states:

> "The test shows that no advantage is obtained by the Calvert pattern, while the additional pressure due to the resistance in the pipes of the Economiser under trial, and the objectional liability to trouble arising from the stoppage of the scouring valves, constitute important disadvantages inseparable from this Economiser."

MICHAEL LONGRIDGE, M. Inst. C.E., Chief Engineer of the Engine, Boiler, and Employers' Liability Insurance Company, Limited, Manchester, in his report, dated September 11th, remarks on the tests as follows:—

"They prove conclusively that under the conditions under which the experiments were made, which are the conditions under which Economisers are usually worked, no appreciable advantage is derived from Circulation, whether the Economiser be Green's or Calvert's, while the liability to short circuiting, which appears to exist in the Patent Circulation Economiser, is a decided disadvantage."

LAVINGTON E. FLETCHER, M. Inst. C.E., Chief Engineer of the Manchester Steam Users Association states:

"The figures obtained during the tests made September 14th 1894, do not show any practical advantage in the continuous current system over the ordinary system."

EDWARD G. HILLER, M. I. Mech. E., Chief Engineer of the National Boiler Insurance Company, Limited, Manchester, remarks as follows with regard to three separate tests that he conducted on June 1st, June 8th, and September 13th 1894:

"Taking the three tests into consideration, a little variation of the results in detail is to be observed, but generally the deviations do not exceed those which might be expected had the same apparatus been tested on each occasion, and considered as a whole they do not show any material economical difference between Green's Ordinary Economiser, Green's Circulating Economiser, and Calvert's Economiser."

CIRCULATORS *versus* ECONOMISERS

HE success of Green's Economiser, as already remarked, has at various times given rise to various imitations, and with a view to securing as inventors have fondly imagined greater efficiency, designs have been put forward which indicate a complete misconception of the distinctive feature of Mr. Edward Green's first invention. This is most clearly shown in the types of so-called "feed-water heaters" which are placed in the down-take at the back end of the boiler with a view to obtain the advantage of high temperature, and which is illustrated in Mr. Twibill's Patent, No. 2,378, 1866, described on page 38.

In this design the "heater" is coupled to the boiler, both at the feed and outlet end, and is really a circulator, and as far as the economy of the boiler is concerned, is merely an extension of the boiler-heating surface. As inventors do not appear to grasp the fallacy which underlies this type of heater, and as numerous designs embodying the same principle have at various times been patented, it may be of service to steam users to point out why such designs, as compared with the Green's Economiser, must from the nature of the case be foredoomed to failure.

It is obvious that in any boiler the temperature of the waste gases cannot be reduced below that of the water and steam. Any attempt to attain this end by extending the heating surface would result in the boiler heating the gases instead of the gases heating the boiler, and thus prove a source of loss. As a matter of fact, it is found in practice that the temperature of the gases can only be brought down to within about 100° F. or 150° F. of this point. Thus in a boiler working say at a pressure of 100 pounds, corresponding to a temperature of about 340° F., the waste gases must necessarily pass away at a temperature of 450° to 500° F., and provided, therefore, the heating surface of the boiler is sufficient to bring the temperature of the gases down to this point, no possible advantage can accrue from a mere extension of the capacity and heating surface of the boiler itself. Now, in the majority of cases, the proportions of the boiler are sufficient to effect this, and therefore the addition of a "Circulator" of the type shown can produce no economy whatever. To call such an apparatus an "Economiser" in the sense in which the word is now generally understood—*i.e.*, as a separate and distinct vessel for heating the feed-water with waste heat—is misleading.

Compared with such an apparatus, the "Fuel Economiser" embodies an entirely distinct principle. It is, in fact, a separate boiler working within lower limits than are available with the ordinary steam generator, and this it was that constituted the distinctive feature of Mr. Edward Green's first invention. To illustrate the matter in another way: It may be stated that, with an ordinary steam boiler and chimney draught, it is only possible to lower the temperature of the gases from that of the furnace, about 1,600° F., down to that of the flue outlet, say 500° F. As already pointed out, this latter temperature cannot, as a rule, be reduced to any serious extent by means of the boiler alone. With the aid of Green's Economiser, however, this terminal temperature becomes available as the source of heat for raising the feed-water from about 90° F. to something like 300° F., the rise being accompanied by a fall in the temperature of the gases, which without the aid of some such apparatus would be impossible.

Green's Early Tubular Boilers

THE great strength, combined with the large extent of heating surface and the facility for easy handling which became available by the introduction of the tubular sectional element in the construction of Green's Economisers, naturally suggested its application for steam generating purposes in other directions besides that for which it was first employed. Mr. Green, in fact, appears to have

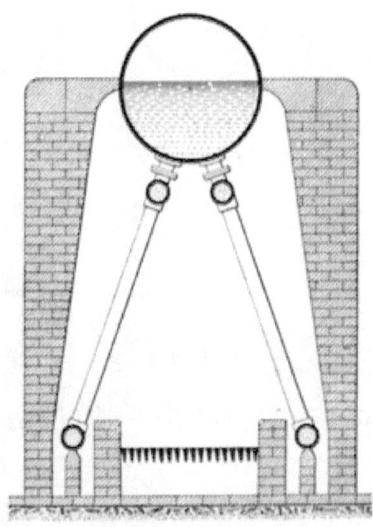

FIG. 66. TUBULAR HIGH-PRESSURE BOILER, CONSTRUCTED BY MR. E. GREEN ABOUT 1850.

realised at a very early date that economy in the application of steam was to be sought in two directions, namely: The extension of the boiler-heating surface, and the use of cylindrical vessels of small diameter with a view to the use of higher pressures. Some of the designs suggested and executed by him bear such a striking resemblance in many respects to the forms of sectional boiler in

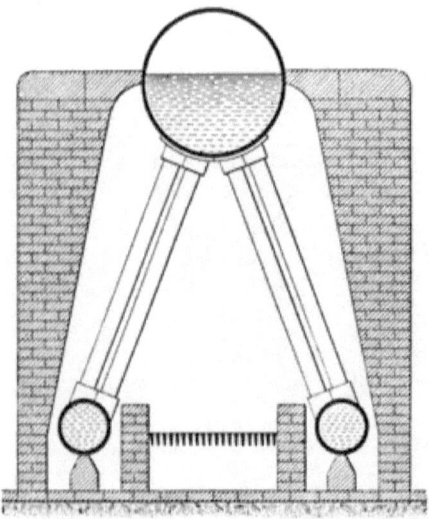

FIG. 67. TUBULAR HIGH-PRESSURE BOILER, CONSTRUCTED BY MR. F. GREEN ABOUT 1850.

successful use at the present day, that it may not be out of place to refer briefly to them here in view of the claims for novelty that are sometimes put forward in connection with boilers of this class.

Figs. 66 and 67, for example, give cross sections of two designs of boilers of the tubular class which were proposed and constructed by him about the year 1850. The general features of the design are in many respects admirable, and as a vessel

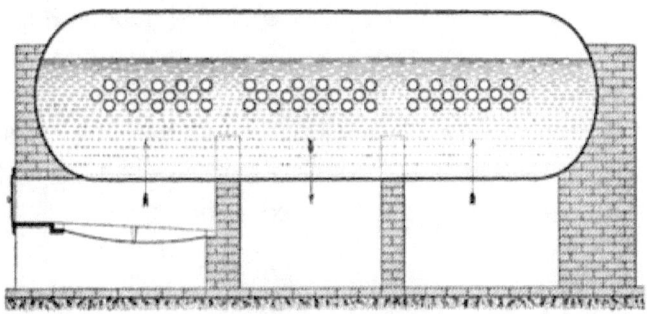

FIG. 68. MR. E. GREEN'S MULTITUBULAR EGG-ENDED BOILER, 1853.

for the generation of steam at high pressures, a striking advance on most of the ideas that had up to that time been proposed. The boiler, it will be seen, consisted of an upper drum connected to two lower longitudinal water cylinders by one or two rows of tubes, which formed inclined walls of heating surface on which the fire played.

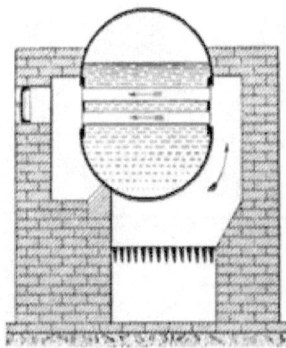

FIG. 69. CROSS SECTION OF BOILER SHOWN IN FIG. 68.

**E. Green
Patent No. 2,882.
1853**
Illustrations, Nos. 68 to 70, represent two of several designs embodied by Mr. Green in Patent No. 2,882, 1853, his object being to increase the heating surface in the plain externally-fired boilers, which were the current type at that period. The idea is so clearly set forth in the illustrations shown that little description is necessary. The boiler, instead of being cylindrical, was made somewhat oval in

section, with flat sides forming tube plates, having several horizontal rows of tubes. These afforded a very considerable extension of the heating surface, and very much increased the efficiency of the plain, cylindrical boiler. The tubes, as will be seen from Fig. 68, were arranged in several groups, the heated gases passing through these in succession, from side to side, until the back end was

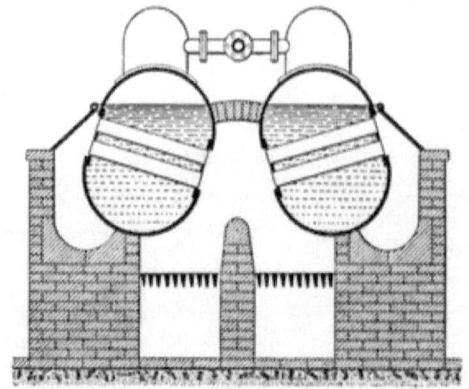

FIG. 70. PAIR OF MR. E. GREEN'S EGG-ENDED MULTITUBULAR BOILERS.

reached, when they were discharged into the chimney. Fig. 70 is a slight modification of the design, showing a pair of this type of boiler working side by side in conjunction.

E. Green and E. Green, Jr. Patent No. 2,671, 1861

In 1861, the first Mr. Edward Green, in conjunction with his son, then Mr., now Sir Edward Green, took out a patent, No. 2,671, for a variety of applications of the sectional principle of construction, not only to the design of steam boilers, but also to condensers and superheaters. Space will not permit of more than a brief reference to these. It may be stated, however, that many of the designs shown in this early specification

—and to which those interested in the subject may be referred for fuller information—anticipate in principle many of the improvements that have been adopted in these kinds of apparatus during recent years.

Figs. 71 and 72 for instance show the plan and elevation of a Sectional Tubular Boiler, with tubes of tapered polygonal or fluted section. The fire was arranged in an external brick furnace,

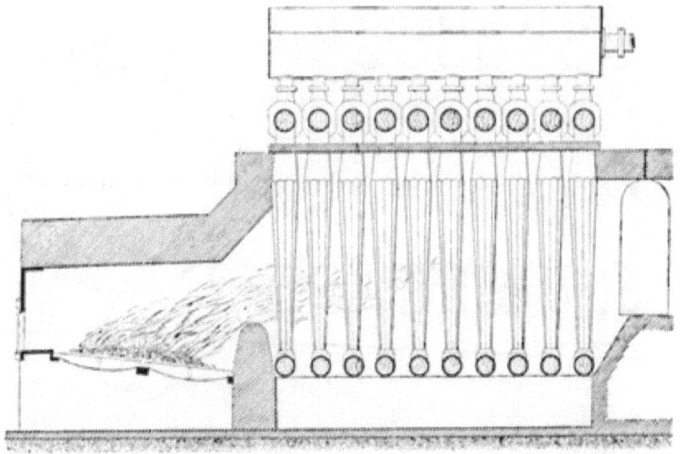

FIG. 71. GREEN'S SECTIONAL BOILER WITH EXTERNAL FURNACE, 1861.

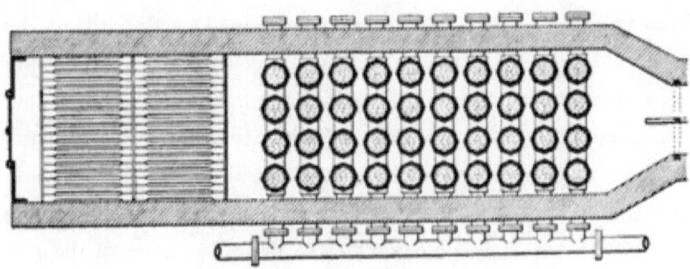

FIG. 72. PLAN OF BOILER SHOWN IN FIG. 71.

whilst vertical pipes were coupled at the top to a collecting steam drum. In some of the other designs proposed, the furnace was surrounded with a water jacket or with walls of vertical tubes in close contact with each other, resembling in this respect the design of several of the Water Tube Boilers adopted in current practice.

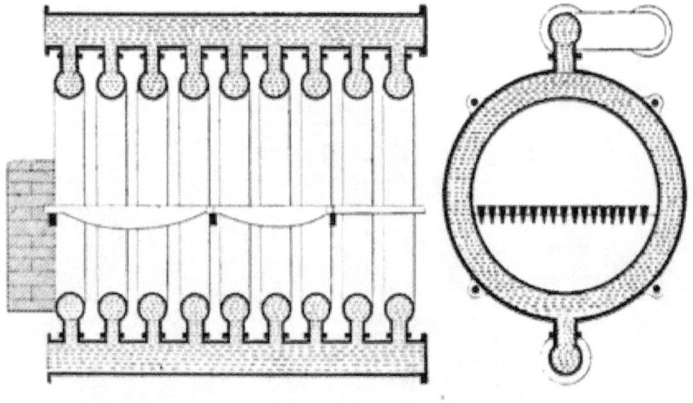

FIG. 73. GREEN'S SECTIONAL FURNACE, 1861.

Fig. 73 shows another proposed arrangement in which the furnace was circular, and consisted of hollow rings built up on a common axis, so that when a number were placed together the furnace assumed the form of a tube. This design for the furnace was proposed with a view to prevent the loss of heat by radiation that is apt to attend the working of an external brickwork furnace, and which sometimes militates against its efficiency. The arrangement shown in the sketch illustrates, of course, only the furnace portion of the boiler exposed to the direct radiation of the fire, the remaining portion of the heating surface, as well as of the

GREEN'S EARLY BOILERS

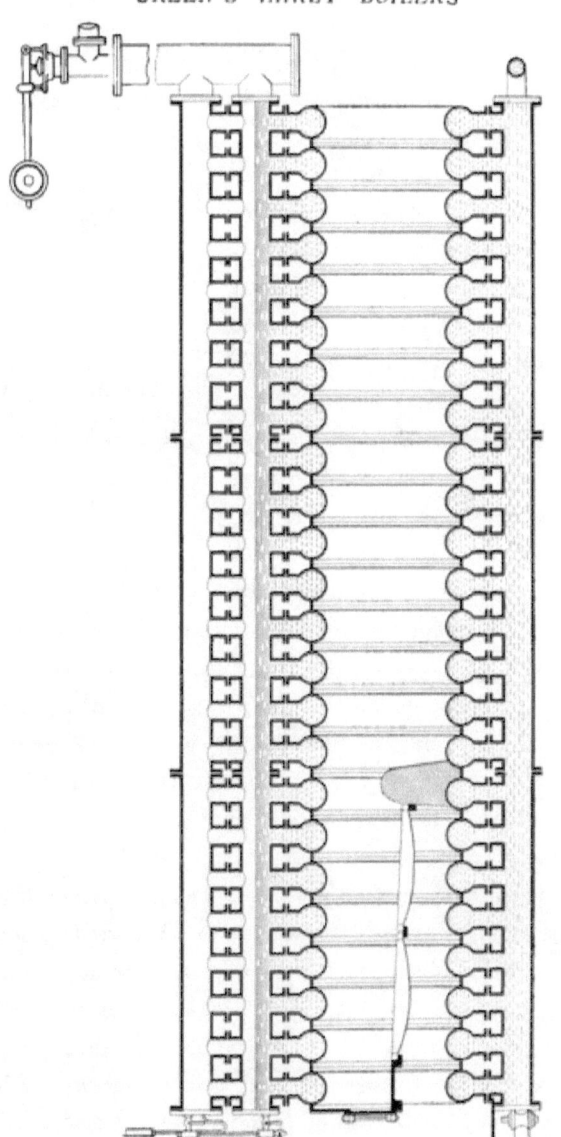

FIG. 74. GREEN'S SECTIONAL FURNACE BOILER, 1860.

steam-collecting drum, permitting of being arranged, either as shown in Fig. 71 or in accordance with one of the many other designs described in the Specification.

Fig. 74 shows a sectional boiler consisting practically of a long furnace tube built up of a series of rings, such as have been described, and which was made by Messrs. Green in order to supply steam for driving their own works in the year 1860. The boiler was practically of the Cornish type, and was set in brickwork with external flues, so as to utilise the heat of the gases and prevent radiation. The rings of which the furnace tube was composed were connected with a water-pipe at the top, above which was placed the steam-collecting pipe, as shown in Fig. 75. As a sectional boiler for high pressures the design possesses many excellent features, and was much superior to the majority of designs that had up to that time been proposed. The boiler proved very satisfactory in practice, and did duty at Messrs. Green's works for many years.

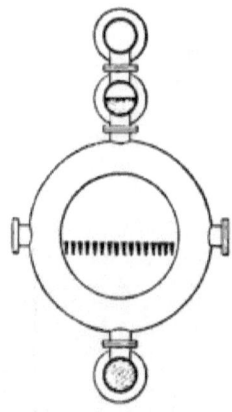

FIG. 75. CROSS SECTION OF BOILER SHOWN IN FIG. 74.

Fig. 76 shows the manner in which a number of Economisers were connected together and fitted with a steam drum, so as to form a steam boiler fired with the flames from an external furnace. This arrangement was first adopted to meet the wants of a sugar works at Greenock, about the year 1863, and was subsequently applied in a number of other cases with very satisfactory results.

Figs. 77 and 78 show two other designs of sectional tubular boilers constructed by Messrs. Green during the years 1862 and

GREEN'S EARLY BOILERS

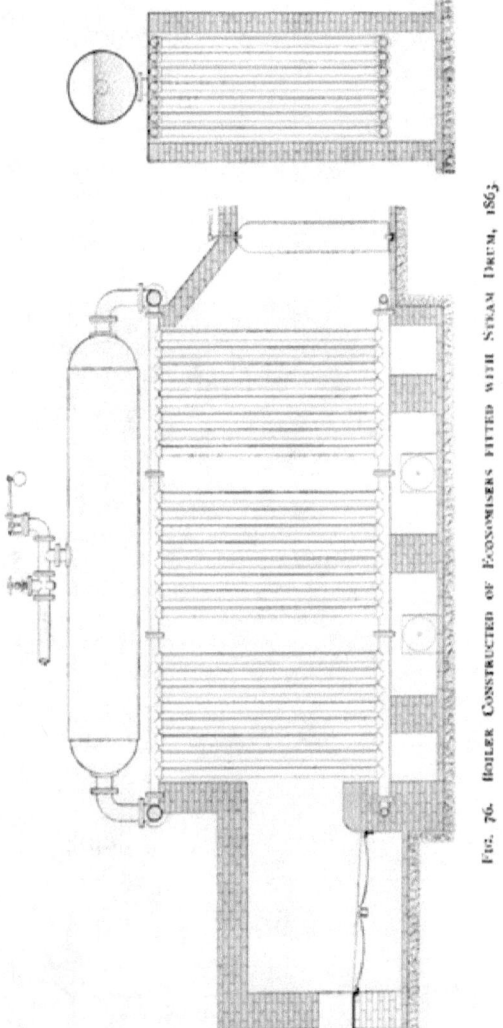

FIG. 76. BOILER CONSTRUCTED OF ECONOMISERS FITTED WITH STEAM DRUM, 1863.

86 GREEN'S EARLY BOILERS

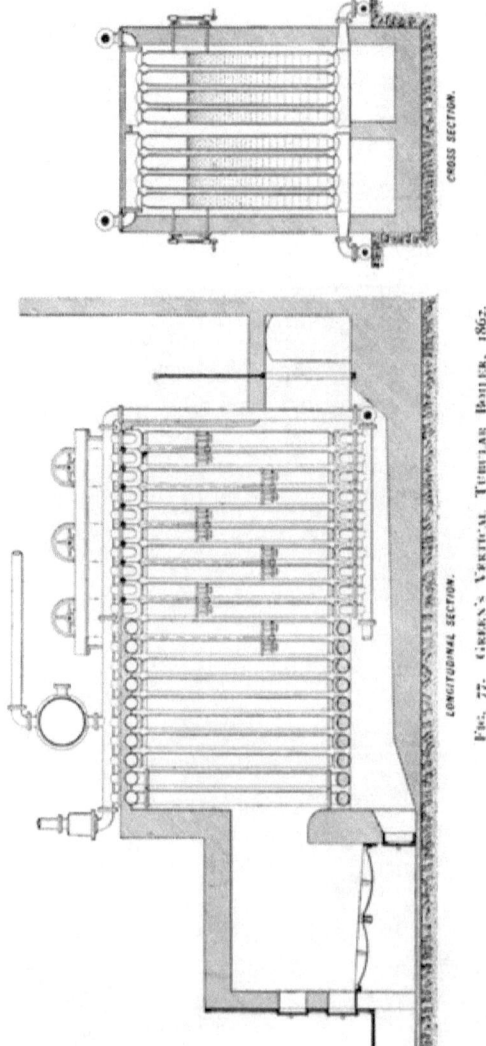

FIG. 77. GREEN'S VERTICAL TUBULAR BOILER, 1862.

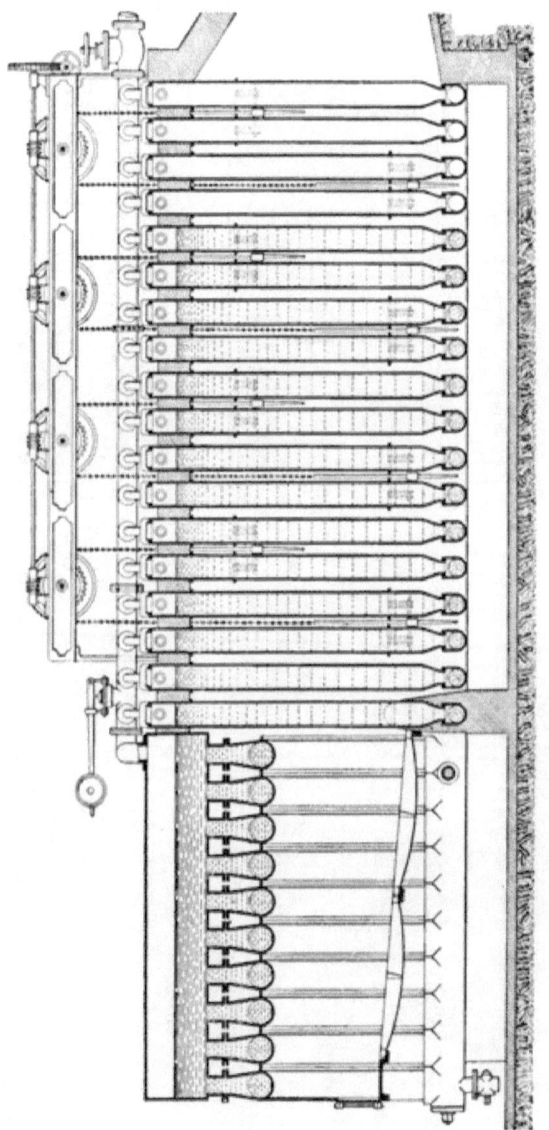

FIG. 78. GREEN'S VERTICAL TUBULAR BOILER FITTED WITH SECTIONAL FURNACE, 1862.

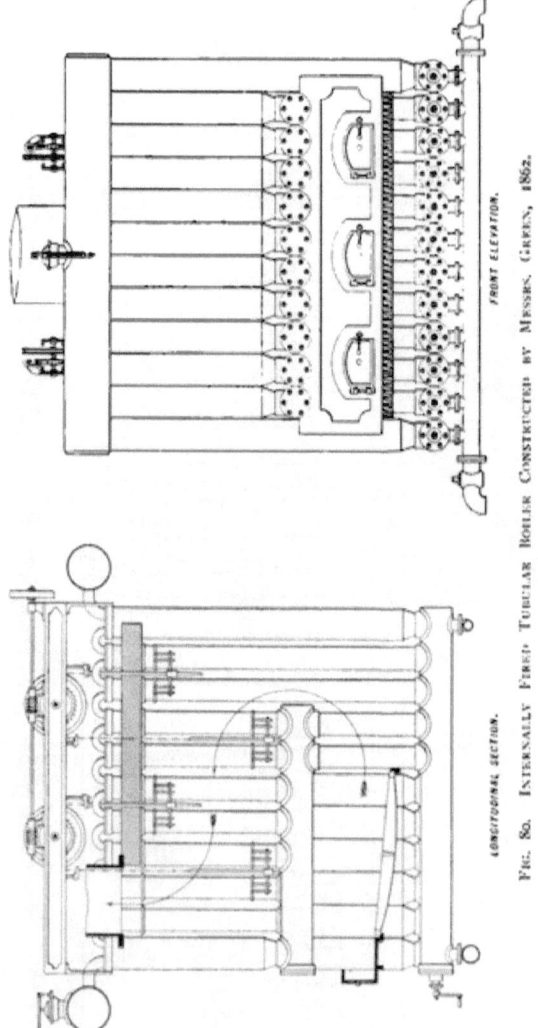

Fig. 80. Internally Fired Tubular Boiler Constructed by Messrs. Green, 1862.

1863. In the design shown in Fig. 78 the furnace was constructed in sections somewhat similar to those shown in Figs. 73 and 74, except that the sections were of horse-shoe shape, as shown in Fig. 79, instead of being circular. This permitted of the grate being lowered if required, so as to afford a large furnace capacity, without diminishing the grate area. This is a point of considerable advantage in the burning of wood fuel, and a number of boilers of this type were sent out to Russia and worked for upwards of 20 years with very good results.

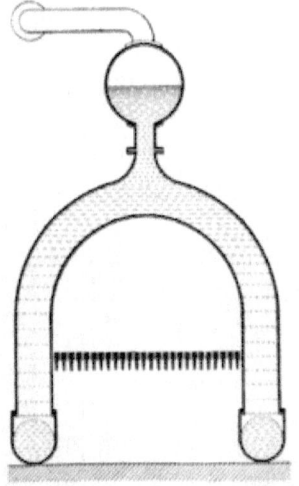

FIG. 79. SECTION THROUGH FURNACE OF BOILER SHOWN IN FIG. 78.

Fig. 80 shows another compact design of tubular boiler, of which a number were constructed about the year 1862. The sides and back as well as the top and bottom of the furnace were composed of water tubes, thus securing the economic advantages of internal firing. The boiler was one which afforded a large amount of heating surface in a relatively small space, and with good water worked with very satisfactory results. For the purpose of facilitating the removal of any sediment which was contained in the water, and which naturally settled in the lower row of tubes, the latter were fitted with mechanical agitators. By means of a handle these could be revolved and the accumulated mud periodically discharged through the blow-out tap.

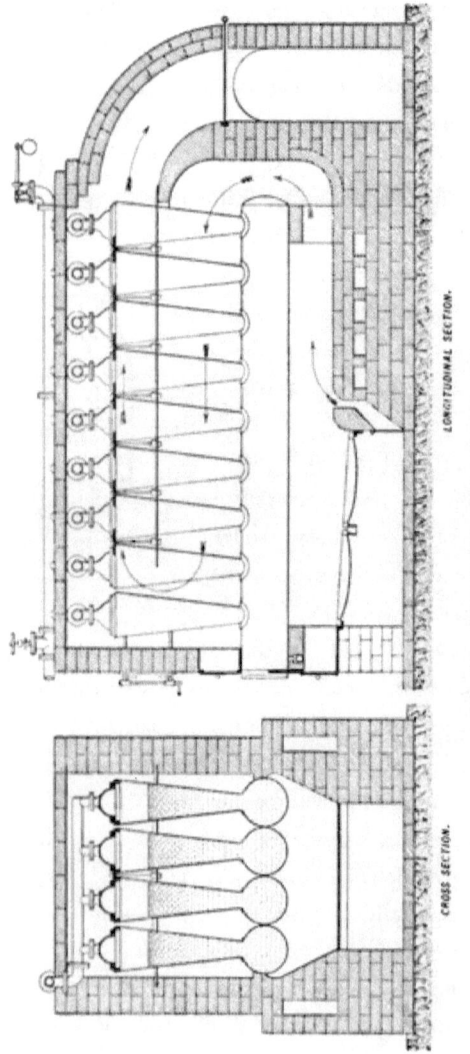

FIG. 51. GREEN'S CONE TUBE BOILER, 1870.

GREEN'S EARLY BOILERS

E. Green Patent No. 1,416, 1870

Patent No. 1,416, 1870, refers to another form of Sectional Boiler proposed and constructed by Mr. Green. The boiler, which is shown in Fig. 81, consisted of a series of horizontal drums lying side by side in contact with each other, so that the lower half of their circumference formed the roof of the furnace. Each of these drums was connected with a row of tapered vertical tubes, the upper parts of which formed the steam space, as shown in the cross section. The flames, after passing along the underside of the lower drums, were deflected towards the front by means of a baffle plate through the nest of vertical tapered tubes, and thence returned to the chimney, as shown by the arrows. The tubes in each row were connected together in the steam space by short necks, while the whole of the rows were coupled to a common main steam-pipe. The tapered form of tubes afforded easy access between the rows to a man for the purpose of inspection, while their upper ends were fitted with removable covers, so that the interiors could be examined when desired.

Green's Early Superheaters

AS the influence of steam superheating in promoting economy in the working of steam engines is a question in which interest has been somewhat revived during recent years, it may not be out of place to note that the subject was one to which Messrs. Green devoted considerable attention at a com-

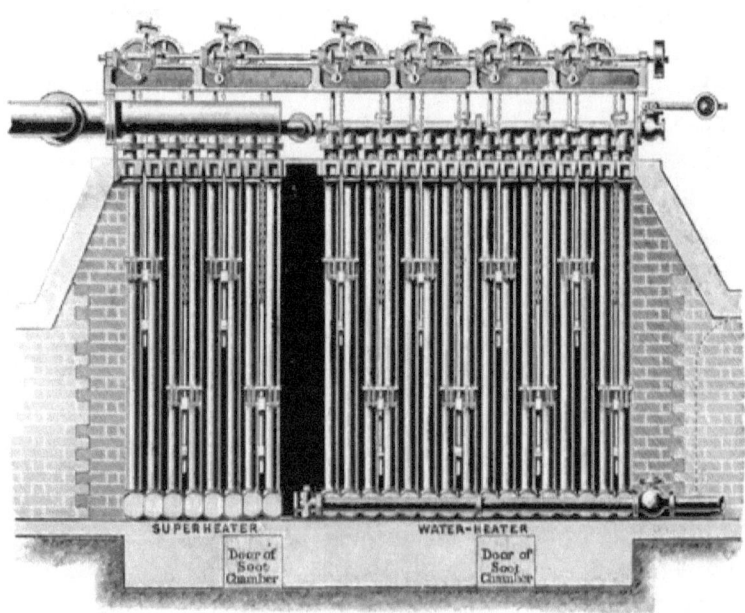

FIG. 82. GREEN'S SUPERHEATER, 1858.

paratively early period. The accompanying illustration, Fig. 82 shows a superheating arrangement which was erected by them in the year 1858. The superheater, it will be seen, was constructed on the same sectional principle as the Economiser, and in fact placed in the same chamber, being situated however at the inlet end, and thus exposed to the highest temperature of the waste gases, which were afterwards used for heating the feed water in the tubes of the Economiser proper. A number of superheaters of this kind were erected, and where circumstances were favourable proved satisfactory.

The adoption of superheating, however, forty years ago was attended with a number of practical difficulties that frequently outweighed its theoretical advantages. The high temperature of the steam gave rise to considerable trouble in connection with the lubrication of the cylinder walls and valve faces, as well as the packing of the glands. The former were frequently scored and abraded, and the latter rapidly burnt out. Since then great advances have been made in the manufacture of special lubricants for surfaces working at high temperatures as well as in the construction of metallic packings for pistons and glands. By these means the troubles which were formerly experienced in connection with the use of highly-superheated steam, and which eventually led to its disuse, have been largely overcome. It will be seen, however, that the superheating of steam is not the novelty that is sometimes supposed, and that in connection with its first introduction the firm of Messrs. Green played a not unimportant part.

Green's Modern Economiser

HE accompanying illustrations represent Messrs. Green's improved arrangement of Economiser as developed by the process of evolution sketched in the preceding pages, and which may be said to fairly embody all the good features suggested by special study combined with fifty years' practical experience in its manufacture and working.

The apparatus, as will be seen on reference to the sketches on the opposite page, consists of a set of cast-iron tubes about four inches in diameter by nine feet in length, made in sections of various widths to suit the convenience of the boiler plant, and set vertically in parallel rows so as to afford facility for inspection. These sections are connected together by transverse pipes or boxes at the top and bottom, which are in turn connected to branch pipes running lengthwise on opposite sides, and situated outside the brickwork which encloses the arrangement.

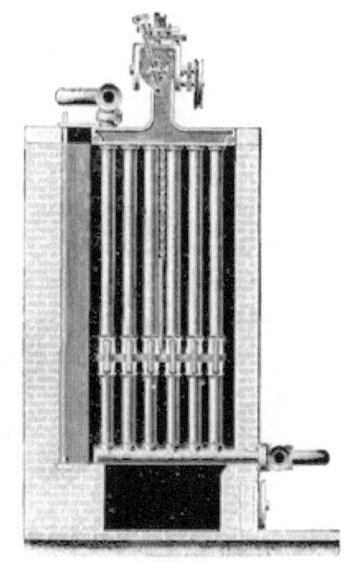

FIG. 85. CROSS SECTION THROUGH ECONOMISER CHAMBER.

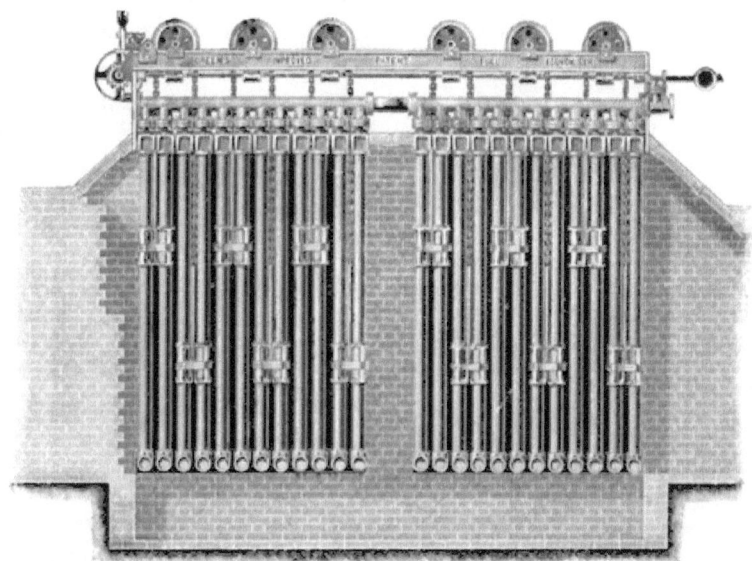

FIG. 83. LONGITUDINAL SECTION THROUGH ECONOMISER CHAMBER.

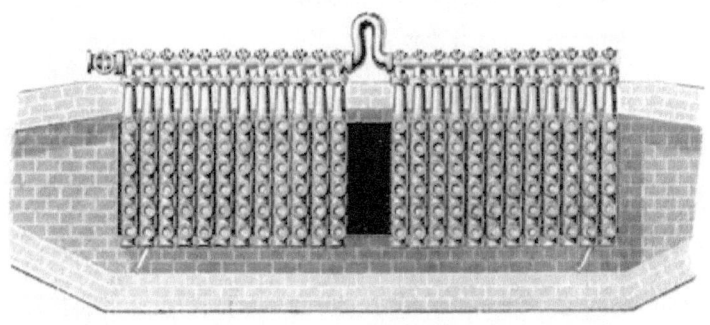

FIG. 84. PLAN OF ECONOMISER CHAMBER SHOWING PASSAGE FOR INSPECTION.

The apparatus is fixed in a bye-pass situated in the main flue between the boilers and the chimney, and fitted with suitable dampers so that it may be at once isolated for the purpose of inspection or repair without in any way interfering with the working of the boilers.

Along one side of the Economiser a passage is generally left wide enough to admit of a man for the purpose of inspection, while underneath the bottom boxes is a chamber for the collection of the soot removed by the scrapers, provided with suitable doors to admit of periodical cleaning.

The feed-water is forced into the Economiser through the bottom branch pipe at the end of the apparatus where the gases make their exit. From this branch pipe it has free access to all the bottom boxes, and rising slowly through the whole nest of tubes makes its escape from the top branch pipe at the opposite end of the Economiser where the gases enter. Thus **the water inlet end** of the Economiser **is the gas outlet end, and the water outlet the gas inlet end.**

At the end of the bottom branch pipe opposite to the feed inlet a blow-off valve is fixed, so that any accumulated mud or sediment which collects in the bottom boxes can be blown out as required. At the end of the top branch pipe opposite to the outlet valve a safety valve is fixed to prevent any excessive accumulation of pressure.

Each tube is provided with a scraper, which is made to travel continuously up and down the tube at a slow rate of speed, so as to keep the external surface free from any non-conducting coating of soot, and thus maintain its efficiency. These scrapers are grouped together, so that those on two adjoining sections of tubes form a single set, which is balanced by the set on the next two sections of tubes. The scrapers are thus exactly counterpoised, and very little power is required for their operation.

The driving of the scrapers is effected by means of a shaft running the length of the Economiser, and carrying worms at intervals which gear into toothed wheels attached to the chain

FIG. 86. GREEN'S QUICK REVERSING MOTION FOR SCRAPERS.

pulleys, over which the scrapers are suspended. At one end of the Economiser is an automatic arrangement by which the motion of the scrapers is reversed as they alternately reach the top and bottom end of the tubes. See Fig. 86. The motive-power for

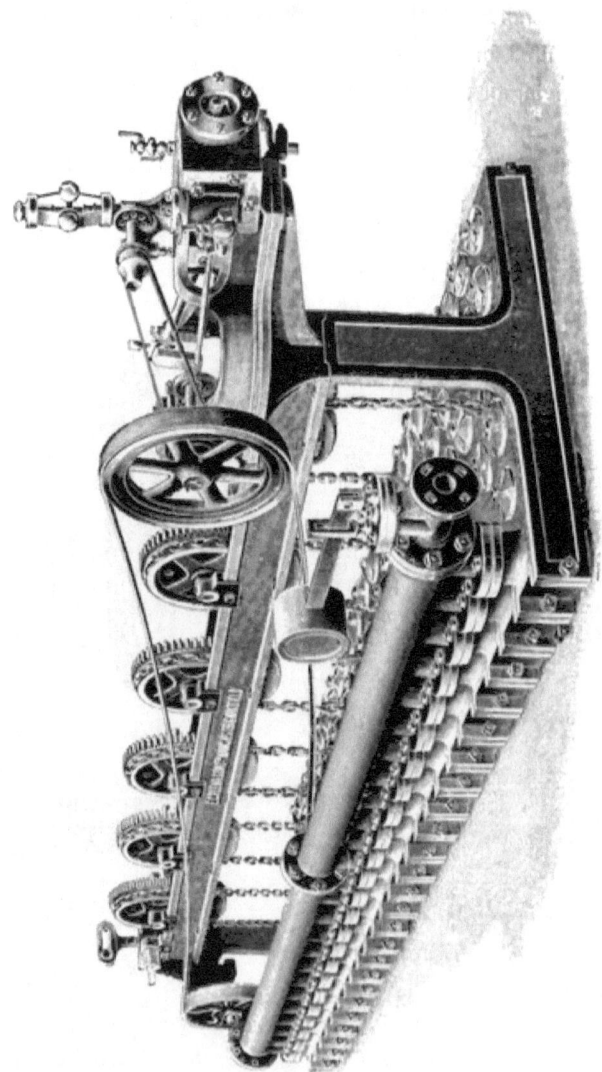

Fig. 87. View of Top of Economiser showing Independant Engine for Driving Scrapers.

the driving may be supplied either by a belt from some convenient shaft or by a small separate engine, as shown in the illustration on page 98.

The tubes of the Economiser are made from a special combination of best Scotch pig and Hematite, cast vertically in dry sand moulds, and are guaranteed to be of equal thickness as well as sound and free from blow holes throughout. Every tube used is tested before being sent out of the works by hydraulic pressure to 650 pounds on the square inch.

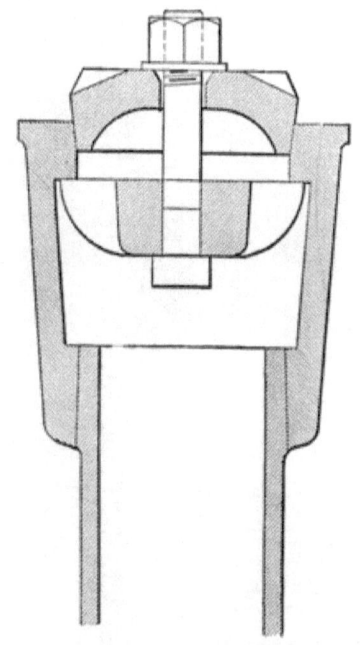

FIG. 88. EXTERNAL LID FOR ORDINARY PRESSURES.

Equal care is taken in the construction of the top and bottom boxes. The joints of the tubes and boxes are all conical, turned and bored metal to metal, and forced together by powerful hydraulic machinery specially designed for the purpose. The lids and holes to each tube are also turned and bored conical to template, so that the joint can be made without the intervention of any hemp or red lead. See Figs. 88 and 89.

The scrapers are carefully designed so as to readily accommodate themselves to the tube, and have a thin cutting edge so as to prevent any accumulation of soot. See Fig. 12, page 21.

All parts of the Economiser are made strictly to template on the interchangeable system, while duplicate standard castings of every part are always in stock, so that any portion can be renewed at a moment's notice.

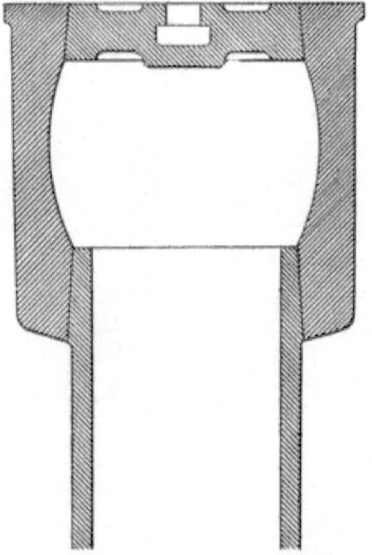

FIG. 89. INTERNAL LID FOR HIGH PRESSURES
(TRIPLE AND QUADRUPLE EXPANSION.)

It is of interest to note that the apparatus can be installed without any stoppage of the works or interference with the working of the boilers; and in the case of a tube failing, an event which rarely occurs, it can be replaced in an hour or so by unscrewing one nut and lifting one lid. See explanatory illustration, Fig. 90.

FIG. 90. METHOD OF WITHDRAWING DAMAGED TUBE BY MEANS OF LEWIS WEDGE AND DRAW BOLT.
1. Tube as ordinarily fixed.
2. Tube Slit at top to permit of being sprung through hole.
3. Draw-bolt and Lewis Wedge in position.
4. New Tube in position.
5. Stop Ferrules for isolating damaged tube

Size and Efficiency
of
Green's Modern Economiser

THE economy effected by the adoption of a Green's Economiser varies from 10 to 20 per cent. of the total fuel consumed, according to the temperature of the escaping gases from the boiler.

The temperature of the gases is reduced, on an average, from 650° F. on the boiler side of the Economiser to 350° F. on the chimney side, while the temperature of the feed-water is increased, on an average, from 180° to 200° F.

It is not desirable under ordinary circumstances to cool the gases below 350° F., as this temperature is generally required to maintain the natural draught in the chimney. If the temperature of the gases on leaving the boiler does not exceed 400° F. an Economiser is not as a rule advisable.

The feed-water should be supplied to the Economiser at as high a temperature as possible, in order to prevent the watery vapour, in the products of combustion, condensing on the outside of the cold tubes. Such "sweating" or condensation is apt to cause external corrosion. If the feed-water is under 90° F. it is a good plan to connect a ¾-inch pipe from the hot water outlet of the Economiser to the suction pipe of the pump, in order to take off the chill and prevent the risk of "sweating."

The percentage of gain resulting from the increase of temperature of the feed-water may be found by the following formula:—

$$\text{Gain per cent.} = \frac{100\,(T-t)}{H-t}$$

Where H = Total heat of Steam at Boiler Pressure reckoned from 0 Fahrenheit.
,, t = Temperature of Feed-water in degrees Fahrenheit, before heating.
,, T = Temperature of Feed-water in degrees Fahrenheit, after heating.

The following clear widths should be allowed inside the Economiser Chamber for various sizes of Green's Economiser:—

				ft.	in.		
Economiser	4 tubes in width	3	4	inside chamber.
,,	6	,,	,,	4	8	,,	,,
,,	8	,,	,,	6	0	,,	,,
,,	10	,,	,,	7	4	,,	,,

NOTE.—The above widths are exclusive of that necessary when a passage is allowed down the side of the Economiser for the purpose of inspection, and for which an additional 9 inches should be added to the dimensions given above.

In fixing on the size of an Economiser for a given boiler plant, four tubes should be allowed for each ton of coal consumed per week. Thus, if 20 tons of coal are consumed per week, the Economiser should contain not less than 20 × 4 = 80 tubes. If the coal consumption is not readily ascertainable, another convenient rule is to allow one tube for every three Indicated Horse-Power. Thus, 300 I.H.-P. would require about 300 ÷ 3 = 100 tubes.

On account of the scraper arrangements the number of tubes in an Economiser must be increased or diminished by not less than four sections. Thus, if the Economiser is four tubes in width, its size requires to be altered 4 × 4 = 16 tubes at a time. If it is six tubes in width the alteration in size is 6 × 4 = 24 tubes at a time; and so on.

If the number of tubes in an Economiser exceeds 96, it is better to divide the Economiser into one or more groups, with

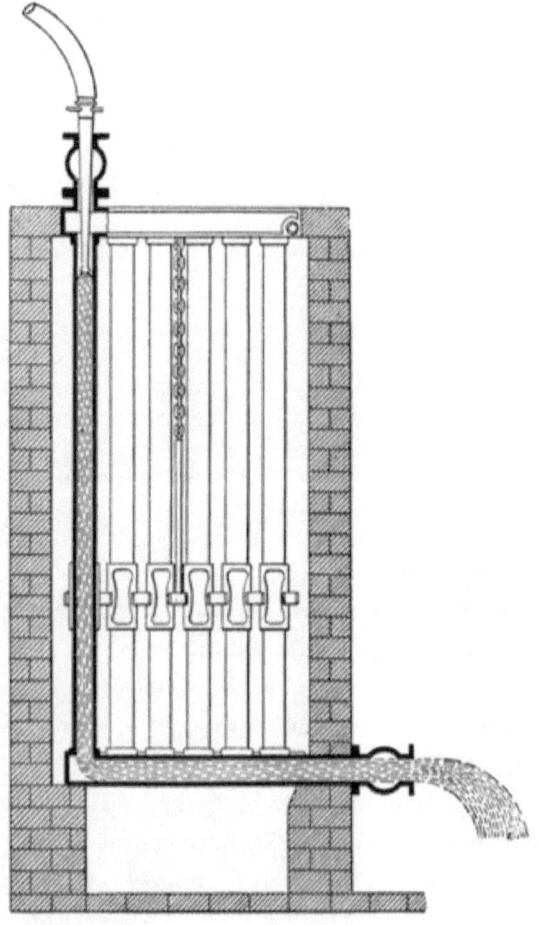

FIG. 91. METHOD OF FLUSHING BOTTOM BOXES WITH THE AID OF ACCESS LIDS.

space between each group to permit of inspection, the top and bottom branch pipes of the several groups being connected together by suitable bends, so as to allow freedom for expansion and contraction. See Figs. 83 and 84, page 95.

FIG. 92. ACCESS LIDS FOR TOP BOXES TO PERMIT OF FLUSHING AS SHOWN IN FIG. 91.

Each Economiser tube holds on an average about 5 gallons of water, including top and bottom boxes. Hence the number of tubes × 6 gives the capacity of the Economiser in gallons. Thus the capacity of an Economiser with 96 tubes = 96 × 6 = 576 gallons.

Green's Economiser
Directions for Working

Raising Steam — When raising steam the reserve flue damper should be kept open until the Engines are started.

Connections to Boilers — The Outlet Valve between the Economiser and the boilers should not be closed when raising steam, or during the night, or at meal times.

Regulation and Temperature of Feed — Feed the boilers constantly, keep the boiler feed valves open and regulate the feed, as far as practicable, by the inlet valve to the Economiser. The Economiser must not be fed with water at a lower temperature than 90° F.

Pump Relief Valve — A Relief Valve, weighted slightly in excess of the working pressure of the boilers, should be placed on the feed pipe, between the pump and the Economiser.

Regulation of Dampers — Regulate the draught of the boiler by the main damper at the outlet end of the Economiser, and not by the boiler dampers. The latter should only be used in case a boiler is laid off.

Leakage of Air — Prevent any cold air leaking into the main flues; make all crevices in the brickwork air-tight.

Blow-out Valve — Test the Blow-out Valve at the bottom of the Economiser daily.

Safety Valve — Test the Safety Valve on the Economiser daily. On large Economisers, or those working at high pressures, an additional direct loaded spring valve may be applied.

Inspection of Valves
All Valves, and especially the safety valve, should be carefully examined periodically, and all waste water delivered so that it will not drain into the soot pit, flues, or cleaning-out space.

Scrapers and Gearing
Keep the Scrapers constantly at work when heat is passing through the Economiser. The cross shaft carrying clutch box should run at 55 revolutions per minute. If the scrapers stick, remove the set screw connecting chain pulley and worm wheel, and work the scrapers up and down by hand until they run freely. The chains and chain wheels must not be allowed to become greasy or they will slip. After a time the chains stretch slightly, and should be shortened. When they get worn turn them round on the pulleys.

Thermometer and Pressure Gauge
A Tell-tale Pressure Gauge and also Thermometer should be placed on the water-outlet pipe, and arranged, if possible, so as to be visible in the boiler-house. In noting the temperature care must be taken to see that there is sufficient mercury in the cup to immerse the bulb.

Non conducting Covering
For covering the top of the Economiser, silicate cotton, slag wool, hair felt, or asbestos only must be used. The branch pipe joints, where possible, should always be accessible for examination, and the Economiser protected by a light roof.

Treatment for Incrustation
If the feed-water be impregnated with lime salts or other impurities, and composition be used, it should be introduced into the feed pipe before entering the Economiser, by means of a small injecting pump.

Cleaning Interior of Tubes
Examine and clean the vertical tubes and bottom boxes internally at least once every twelve months.

Cleaning Soot Chamber Do not allow the Soot Chamber below the Economiser to get too full. It should be cleaned out once a month. The spaces between the vertical tubes, bottom boxes, and side walls should also be thoroughly cleaned.

Accumulation of Soot on Tubes To remove an accumulation of soot from the tubes, shut down and empty the Economiser. Disconnect inlet and outlet feed pipes. Open all valves and pass the heat through the Economiser chamber for several hours, keeping the scrapers constantly running while this is being done. **Before re-filling the Economiser be careful to ascertain that the tubes are quite cool.** The soot should not be burnt off the tubes in this way without expert advice.

Dampness in Flues Any Dampness in the flues, foundations, or soot chamber should receive prompt attention.

Shortness of Water If the water in the Economiser gets too low, immediately open the reserve flue damper and close Economiser inlet damper, leaving outlet damper wide open. Also open direct boiler feed valve and close inlet and outlet Economiser feed valves. Remove covers from soot pit manholes and open all doors into Economiser chamber. Do not touch the safety valve until the tubes are cool.

Emptying in Frost Economisers fixed in exposed positions and standing idle should be emptied during frosty weather.

Shutting down To shut down Economiser open reserve flue damper and close inlet and outlet Economiser dampers, also open direct boiler feed valve and close inlet and outlet Economiser feed valve.

HYDRAULIC PRESSES.

Conclusion

THE brief historical review of the development of Green's Economiser that has been given, will, it is thought, show that the most painstaking efforts have been constantly devoted to its improvement. The design and manufacture of every detail have been made the subject of special study, and countless experiments have and are being made, of which it is impossible to take any note in these pages. It is sufficient to say that in its present form the apparatus represents the experience of a firm which has for a generation been devoted to its manufacture and working. Possessing, as Messrs. Green do, such a lengthened and unique experience, they not unnaturally feel confident that they produce at the present day an Economiser which for efficiency is not excelled, and which ought not to be lightly passed over.

www.ingramcontent.com/pod-product-compliance
Lightning Source LLC
Chambersburg PA
CBHW020103170426
43199CB00009B/380